Mandala Urbanism, Landscape, and Ecology

Archana Sharma

Mandala Urbanism, Landscape, and Ecology

Interpreting Classic Indian Texts and Vaastupurusha Mandala as a Framework for Organizing Towns

Archana Sharma
Morgan State University
Baltimore, MD, USA

ISBN 978-3-030-87284-7 ISBN 978-3-030-87285-4 (eBook)
https://doi.org/10.1007/978-3-030-87285-4

This Springer imprint is published by the registered company Springer Nature Switzerland AG
The registered company address is: Gewerbestrasse 11, 6330 Cham, Switzerland

To my parents Shri Subhash Chandra Sharma, Usha Sharma, and Family.

And to my friends, teachers, students and mentors.
Thank you.
For everything.

First two lines of a poem with the power to inspire journey through darkest of times, never give up.

दुष्ट दुर्योधन अन्धेंरे में छिपा, काली गदायें फेंकता,
अर्जुन,उठा गांडीव लो , अब युद्ध की है अनिवार्यता ।

Subhash Chandra Sharma, 2019 Spring

Prologue

An archetypal Hindu temple consists of mahadwar or gopura—the main entrance to the premise; a set of halls or mandapas, with one of those marking the entrance to the temple; ardhamandapa—the transitional porticos or halls; mahamandapa—the main temple hall; and antarala—the transitional space leading to the sanctum sanctorum, the main shrine, which is called garbhagrha. The spatial sequence enfolds the Hindu philosophical focus of moving from outward to inward, leaving the outer, material, physical realm behind in incremental succession through each transitional space, and moving closer to God—either perceived as the Spirit and the Universe or the symbol of spiritual, universal force pervading through the cosmos. A tower is typically built directly above the garbhagrha (Sinha, 1996) to highlight the transition from the ground, upward toward the symbolic heavenly, spiritual realm. The Hindu temple is thus conceptualized as a place for communion with God, "a place," not "the only place". The communion can be one to one without any individual or priest-ordained rituals—through darshana—or complemented with rituals of offerings called archana. Darshana can be understood as an individually scripted contemplative, reflective thinking as well as a communion through viewership. Archana as ordained by priest is more formally Vedic scripted communication with God, which is reinforced through mantras and a sequence of ritualistic actions; the practice by an individual takes a relaxed format. The communion thus allows for individual or collective worship. This is a place to reflect on the word of God or to resolve your personal knots in the mind in presence of the divine energy. This communion with God is not restricted to the confines of the temple construct but the construct allows and creates a space for the specific purpose. The spatial constructs thus become a place that manifests the underlying religious philosophy, cultural beliefs, and thinking, on the identities and purpose of the human and the divine, the brahman and the Brahma, the cosmic entity, and the cosmos.

Mantra is the most popular form of communion with God in temples. Mantra is a hymn or a sloka that is chanted, mostly repeatedly, to render more energy, force, or momentum to the words of prayer. Mantra worship could be performed audibly loudly (vachik), mentally (manasik), and in a gentle whisper tone (upanshu), in presence or absence of a pratima—a sculptural idol form as a representational form of the Divine, or a Yantra—the geometric, diagrammatic pattern, mechanization, representing the symbolic energetic embodiment of the Divine. Mantra and yantra are all mutually reinforcing and not mutually dependent; thus, praying through any singular modality is not incomplete.

Circumambulation or pradakshina from the periphery to the center is at the heart of the Hindu religion, and these can be read in the Mandala diagrams as well. The walls enveloping the temple or the shrine form a guide to the pradakshina path, and are recessed to accommodate parshva and avarana devatas. Kramrisch (1976) observes that, in the rite of circumambulation, the devotees become the perimeter of the temple building.

The circumambulation from the periphery of a temple town to the center of the temple is a pradakshina from the periphery to the center of the source. The self-realization, learning, and lessons being realized correspond with the depth of the seeker and will vary for each.

Acknowledgments

The project moved through different stages of conceptualization to research and writing, with the support of several institutions and people. Morgan State University offered me partial time release to resource primary data through library and field visits. Support from Library of Congress Asian Studies Division Director Jonathan Loar in terms of resourcing of ancient texts and literature on the topic in Sanskrit was instrumental in cross-checking some of the claims on Manasara and Sthapatya Veda related texts as referred to in English texts. Assistance provided by the staff of the Earl S. Richardson Library at Morgan State University, Milton S. Eisenhower Library at Johns Hopkins University, and the Dumbarton Oaks Library and Archives in the Washington D.C. enabled the foundational survey of the extensive range of literature available on Hindu culture, philosophy, and architecture.

The engagement of many of my students in the topic was remarkable. I am incredibly thankful for contributions by Maura Ruth Gormley, Shahrouz Ghani, Matthew Hawkins, Abrar Boghaf, Varun Gupta, Jamie Solomon, and, most significantly, William Chris Schoenster, my graduate research assistant, for his dedicated approach to work on this project and digital illustrations of Mandala. I convey my thanks to Siddharth Madiwale for refreshing my memory on basic mathematical formulations, Deepti Sharma for proofing the highlights, and Aadya Tewari for conferring on cover graphics.

I am grateful to the stalwart thinkers from architecture, landscape, and religious Indic studies, Tom Verebes, Diane Jones Allen, and Pankaj Jain, for their reviews. I am also appreciative of the anonymous peer reviewers and the Springer editorial production team, led by Aaron Schiller and including Henry Rodgers, Herbert Moses, and Gomathi Mohanarangan.

It is a delight to read works of honest investigations conducted from a non-judgmental objective place of a researcher willing to be challenged and taught by the findings, especially when foraying into an unfamiliar cultural ground. In English print, Stella Kramrisch undoubtedly and unflinchingly leads the list for expositions on Hindu temples, architecture, and philosophy to the rest of the world, as if the language and culture she was translating were her own. In terms of translations and cultural interpretations of the ancient Indian text, works of Prasanna Kumar Acharya,

Radha Kumud Mookerji, Madhusudan Amilal Dhaky, Dwijendranath Shukla, Krishna Deva, S. K. Ramachandra Rao, Upendra Mohan, Sarvepalli Radhakrishnan, Ganpati Sastri, Prabhashankar Sompura, Michael Meister, Adam Hardy, Richard Davis, Vibhuti Chakrabarti, and Amita Sinha, amongst others. The book takes inspiration from the scholars in urbanism and landscape, too expansive in their influence and ideology to contain or quote in the small space here but influential, nonetheless.

Contents

Chapter 1
Introduction

India has a strongly developed language design and planning language and principles or sutras from far back in histories, such as Kamikagama, Suprabhedgama, Brihatsamhita, Mandukya Upanishad, Mayamatam, Matsyapurana, and Bhavishyapurana, but primarily from Manasara—the ancient Vedic text attributed to 550 BCE and earlier. These texts and other classic ancient Indian texts, most notably Arthashastra, also known as Kautilyashastra from around fifth BCE, articulate design and planning principles, including the organizational frames of grid plans and precede much of the Western grid planning discourse. The book sheds light on these marginally discussed or known spatial organizational frameworks.

In the context of India itself, newer Hindu temple constructions and retrofits of ancient temple structures are based on the classic design and planning principles under the commonly used lexicon of Vaastu Shastra, where Vaastu refers to Vaastupurusha mandala. However, it is not well established whether design and planning at the temple premise and surrounding town follow the classic Vaastupurusha mandala or any other ancient town planning principles. Are these traditional principles and frameworks still valid and applicable in the current context of climate change, or should the classical design principles and Mandala be revised for current realities and forward-moving agility? Guided by this crucial question, the undertaking of this book is to understand and explain the current design syntax, landscape, urbanism, and ecology centered around Hindu temples, investigate re-integration of ancient planning frameworks in present times, point to the Vaastupurusha mandala as a potential spatial organization framework for the sporadic development of temple towns in India, and advance the discussion on the use of design syntax as a tool to understand and explain design and planning.

There is a substantial body of knowledge published on Hindu temples. Most of it is centered around architecture and akara of the temple (Kramrisch 1976, 1946; Ferguson and Burgess 1910; Brown 1959; Dhaky 1996; Meister and Dhaky 1999; Hardy 2007; Acharya 1927a, b), with a few unraveling the interface of the form and philosophy and spirituality (Kramrisch 1976, 1946; Rao 1997; Shukla 1967). There

A. Sharma, *Mandala Urbanism, Landscape, and Ecology*,
https://doi.org/10.1007/978-3-030-87285-4_1

is, what can be called, a gradually expanding scholarship on urbanism through the lens of what has now been pop-termed as "Vaastu," an abbreviation of the Vaastupurusha mandala (Pandya 1998, 35–38; Sinha 2006, 141; Chakrabarti 2013); most of these look at settlement plan first and temple second or temple as a subset of the town, thus spatial organization outside-in, with some exceptions (Deva 1996; Bharane 2012). The discussion on the urbanism and ecology of the Hindu temple, which takes place in the prakara—the precinct of the temples—and radiates outward into the town or city fabric, remains noticeably thin in volume, thus prompting this study. The inquiry runs parallel to inquiries on the Indian identity in design (Bhatt 2001) and urban ecological sustainability (Rademacher 2018).

The focus is spatial, formal design for implications on current urbanism, landscape, and ecology. The book does not intend to explain, question, or challenge Hinduism, Sanatan dharma, Hindu philosophy, Shiva, or Shivalinga as tomes have been written on the topic, and that is what the undertaking requires. Nor does it claim that the temple, prakaras, towns, and cities were laid out according to the *Vaastupurusha mandala* principles or Mandala blueprints as underlays. The study starts from the position of acceptance of the given palimpsest, with a particular focus on physical, formal design and layout, to reveal the congruence or dissonance of current forms with its historical antecedent and critical design deliberations therefrom.

The study considers Shiva temple towns with Dwadash Jyotirlingas among the broad category of Hindu temples as representative religious worship sites under active usage. The Dwadash Jyotirlingas, with high religious significance in Hindu religion, are an ode to the primordial God Shiva, the first-ever Yogi, the one detached from birth, death, reincarnation, space, time, but "is."

The book has seven key chapters.

The first chapter explains the language to the readers to bring diverse understandings regarding the terminology to a common denominator. The text reviews the semantics of the Hindu temple, *Vaastupurusha mandala*, and the philosophy underlying temple visits. In further subsection, the reader is introduced to the archetype of Hindu temple while gradually orienting them to this book's position with typology as a research method for this investigation.

The second chapter introduces the lexicon of the *Vaastupurusha* Mandala and expounds on the classic town planning principles laid out in ancient Indian texts. This includes an exposition of the town planning ideology of Kautilya, commonly known for his economic and political theories. The narrative is presented through a focused study on *Vaastupurusha mandala, Site and Layout principles; Town planning, Urbanism and Design syntax, Landscape, and Ecology*. The chapter thus presents key classic site and town planning mandalas that will be used as a fundamental formwork for assessing contemporary temple towns in subsequent chapters, namely, Dandaka, Chaturmukha, Nandyavarta, Prastara, Padmaka, Sarvatobhadra, and Svastika.

The third chapter uses nomenclature, typology, and design syntax as additional interpretive frameworks to decode the temple syntax. This becomes a place for regarding the ancient from the present viewpoint. The meaning and design syntax of mandira, devalaya, and puri are explored here. Temple network may not be an unexpected outcome but a designed typology.

The fourth chapter presents a case study investigation through a focus on Shiva temples and corresponding temple towns of Kedarnath in Uttarakhand; Vishvanath in Varanasi, Uttar Pradesh; Vaidyanath in Deoghar, Bihar; Somnath and Nageshwar in Gujarat; Bhimashankar, Ghrishneshwar, and Triambakeshwar in Maharashtra; Mahakaal and Omkareshwar in Madhya Pradesh; Mallikarjunam in Andhra Pradesh; and Rameshwaram in Tamil Nadu. The text is orientative and narrates the *anecdotal history* of the temples. The aural transgenerational passage of the religious significance and associated beliefs with those temples are shared and further deliberated to procure historical clues on the period of first construction or founding of the temple, thus beginning a literature review-based verification of the historicity of the temple sites.

The fifth chapter presents a study of Shiva temple towns in India vis-a-vis Mandala formwork. The chapter presents *layout, urbanism, and landscape* as an abstraction, re-synthesis, and articulation of the classic text as a frame for reading the landscape, urbanism, and ecology of the temple precincts and towns. The chapter addresses the research question on the current landscape, urbanism, and ecology of temple towns. The critical review also begins responding to the second question on the relevance of those principles in present times. In the process of superimposition of traditional mandalas on the forms of temple towns, new compound Mandalas are expressed and proposed for future considerations, such as Khand Prastarit Padmaka and Prastarit Padmaka, both being compounds of Padmaka and Prastara with different formal expression; Mandala Mandal, a network of mandalas; Dandamandit Sarvatobhadra, a combination resulting out of Dandaka puncturing Sarvatobhadra; and PrastaraPakshaDandaka, a compound of side-by-side co-located Prastara and Dandaka.

The sixth chapter considers a combination of classic town and landscape principles and currently popular landscape typologies transposed on Mandala formwork to remedy the imbalances between the built and natural environment. To explore the feasibility of Mandala urbanism discourse for other towns beyond the temple towns, a creative interpretive design exercise is conducted to read the developments in the international context and reveal the corresponding Vaastupurusha mandala for Amman in Jordan, Barcelona in Spain, Brasilia in Brazil, and Washington D.C. in the United States.

The closing chapter lays out the Mandala urbanism approach and principles, algorithms for devising new mandala, as a potential framework to address challenges of present times. In summation, the book shares a re-visioned position of the ancient Indian *Vaastupurusha mandala* conceptualized during a different epoch as a springer of Mandala urbanism with overarching spatial organization possibilities and adaptability to account for changing climate.

Methodology

The book uses historiography and hermeneutics to review the ancient design and town planning principles and transpose them in the present to reveal the implications and potential for resolving emerging concerns of towns.

The driving question is best surmised as co-mingling layers of curiosities such as the following: What are the current types and typologies of the temple towns in India? Do they confirm with the classic town organization principles and Mandalas? Should they? Can the "classic" order inform the "new-age," and how? The inquiry is framed in ancient Indian design, planning principles regarded as the prescient organizational framework. The current context is studied as an overlay, thus pointing to the gaps in the design and planning of two different periods and possibilities of newer fusions or modified versions of the Mandala urbanism for organizing towns.

To keep the study focused, the category of Hindu temples is further narrowed to Shiva temples—an ode to the primordial God Shiva and is considered most significant to Hindus and will be used as case studies for this inquiry. The 12 Jyotirlinga temples based in common religious faith, significance, cultural, and geopolitical context, all adhering to Sanatan dharma or Vedic philosophy, offer a reasonably common ground for inferences on spatial design retrofits.

The concepts explored through this book are:

- Landscape, urbanism, and ecology of temple towns
- Design syntax as an instrument to read the design assembly and component of temple precinct and towns
- Mandala urbanism as a frame for organizing towns and landscape

Accounts of ancient Indian treatises and original Vedic texts on town planning, vegetation, and landscape layout principles from India are used to infer design syntax and assess temple layout and ecology. The thread of the ancient tenets is infused with contemporary ideas on urbanism and landscape.

Primary sources were reviewed and interpreted with the assistance of Sanskrit experts and other authors. This includes Acharya's translations from Bhavishyapurana, Manasara, Kamikagama, and Suprabhedgama. Matsyapurana, Bhavishyapurana, Aparajitapraccha (1927a, b, 1934a, b, c, 1946), and Kramrisch's works are based on Matsyapurana and Mayamatam (1946). Sompura's references from Prasadamanjari (1965) presented a breadth of interpretations with original Sanskrit sources for comprehending the Vaastupurusha Mandala and Hindu temple from an architectural design and planning perspective. Dvijendra Nath Shukla's presentation and comments on Samrangana Sutrdhara and Upendra Mohan Dev Sharma's Sankhya teertha (1936) offered a deeper insight into Hindu philosophy. Writings of Sastri (1919), Dhaky (1996), Deva (1996), Rao (1997), Sinha (1998), Meister and Dhaky (1999), Hardy (2007), and Chakrabarti (2013) were instrumental in the extrapolation of contextual design principles for me, from both the native and foreign perspective. Multiple visits to the Library of Congress and appointments for reviewing the rare manuscript collections were conducted to examine the

Sanskrit texts mentioned above. The intention was to verify the interpretation of design principles in English print with the ancient Sanskrit writings. Although there was a sizeable collection of ancient texts, only a few of those referenced above showed up in this collection. A number of additional, uncommonly available books with older prints than available in other libraries were discovered with images of Sanskrit text from seminal ancient texts, accompanied by an English translation, in the process.

The typology of the city and dialectics between the form and the user have been explored by Lynch (1960) and Calvino (1974), Alexander et al. (1977), Eran Ben-Joseph and David Gordon (2000), Leon Krier (1984, 2006), and Thadhani and Duany (2010), with the aim of finding a system for reading but also organizing a city, although narrated from starkly differing vantage points. The book clarifies mandala or panchmahabhoot that pop up in vernacular Indian architecture discourse for the broader professional practice. The investigation is grounded in the vernacular and works with design syntax, typology, and diagrams, as tools of information editing for formal clarity, with creative liberties, as a method to review Vaastupurusha Mandala as a template for organizing landscape, layout, urbanism, and ecology of the towns. The debate on the form-use, pattern-experience, is not resolved here but used as a conduit to advance discussion on the ancient form—Mandalas to organize present chaos of rapidly expanding towns, and the value of design syntax as a method to decode and re-code towns and cities.

Chapter 2
Ancient Indian Design and Town Planning Principles as a Frame for Case Studies

Vaastupurusha Mandala

The lexicon of Vaastupurusha mandala can be better accessed through the understanding of sub-components of Vaastu, Purusha, and Mandala. Mandala in Sanskrit and Hindi literally means a system or organization or a schematic pattern. The literal translation connects to various meanings: circle, district, zone, territory, grouping, collection, part, whole body, network, a path of heavenly bodies, and circumference. Vaastu can have multiple meanings, such as a thing or matter, a manifested component, and a thing or matter that resides in a place or space (akasha). Vaastu is also a derivative of the Sanskrit word Vastutah, which meant as in fact, de facto, actually, certainly, or essentially. Purusha is the golden light, the energy (that combines with the place) resulting in formal manifestations of prakriti (Radhakrishnan 1995, 1994, 1953, 625). Prakriti literally translates to "nature" and can be considered as the prakatya, vyakta, or revealed and manifested form of Purusha. The quest for revealing or creating the divine order on the earth plane was quite normative in ancient times (Singh Rana 2011; Morley and Renfrew 2010).

The widely disseminated understanding of Vaastu as a site, Purusha as the cosmic/uttama man, and mandala as its essential form where *Vaastu* is essentially a square shape and Mandala is an enclosed polygon, which can be converted into a triangle, hexagon, octagon, and circle of equal area (Kramrisch 1946, Vol 1, pg 21), can be further contemplated, within the context of this understanding. There is an unmanifested or avyakta and aprakutya spirit beyond the manifest, the continuing condition of "existing" as a spirit; the concept is applicable to geographical-spatial scales as contracted as the human body or an object or as expanded as a collective of astronomical systems (Fig. 2.1).

The diagram of the Vaastupurusha mandala as handed down through popular culture, books, and practices is represented in the figure above. The diagram shows the figure of a human body or Purusha transposed in a grid. The grid subsections

A. Sharma, *Mandala Urbanism, Landscape, and Ecology*,
https://doi.org/10.1007/978-3-030-87285-4_2

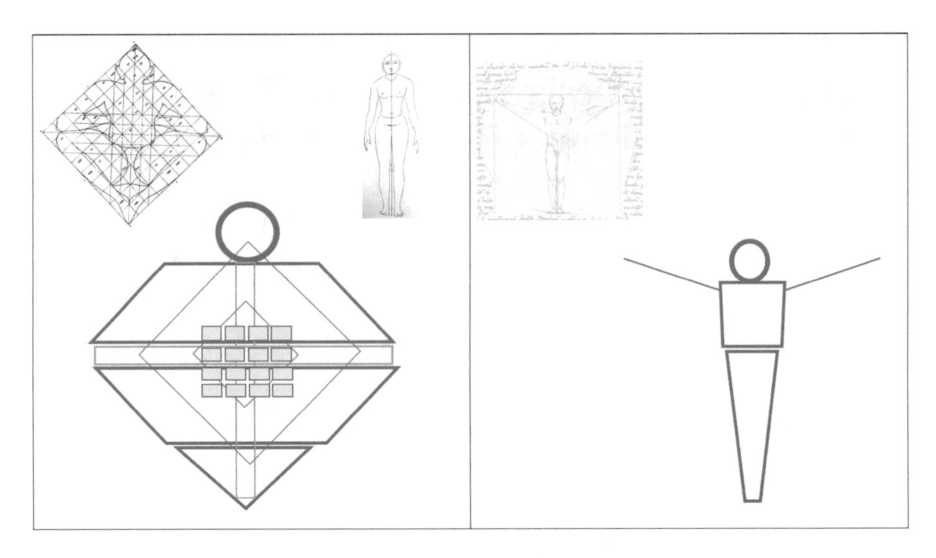

Fig. 2.1 Diagrams of Vaastupurusha Mandala and Vitruvian Man by author: Abstractions inspired by Inset graphic illustration: Vaastupurusha Mandala diagram (top left): Vatsyanan, 1986 in Architexturez South Asia, 2021), Tala-mana (center), (Matsya-Purana, Chap. CCLVIII, v.19 in Acharya 1946, Series 7, pg 196–214), and Vitruvian Man (attribution circa 80–15 BC, source Sgarbi 1993, 41)

and facets correspond with the Sanskrit names of Gods, which are the names of climatic forces of wind, water, fire, and earth, with akasha being space or sky, occupying every interstitial place, and thus not captured in the two-dimensional mandala. The Mandala has as marmasthan or fragile points which should be left empty, and no load-bearing structure should be placed there. For context, the Vaastupurusha mandala, with the attributed diagram extrapolated from Sanskrit slokas, has been discussed at least since 550 BC and earlier. The imagery of Vitruvian Man, often discussed in Western architecture discourse, attributed to Vitruvius, who lived during the era of 80–15 BCE, is represented side by side for comparison based on Sgarbi's print courtesy Biblioteca Ariostea Ferrara. De architectura, Book III, i, 3–7, is where Vitruvius' description for the diagram is given as sacra Aedes or sacred temple (1993, 41). The imagery for the "Tala mana," the measure guide, presented in Acharya based in Matsya-Purana, Chap. CCLVIII, v.19 (Acharya 1923, Series 7, pg 196–214), is a reference used for constructing artifacts ranging from sculptures to buildings. The formulae written in Sanskrit verses on tala-mana are based on human physiology details; the prescriptions are very detailed for crafting nano-micro sculptures to a magnified scale of multi-storied structures and buildings.

Vaastupurusha mandala is thus a schematic system of organization to reveal the essential spirit of the place or to manifest the best form for the same, toward the best outcome for the intended purpose, at any scale. The mandala is a yantra/scheme/machine to be used in the process of revealing, manifesting, and forming the character of the place—the nature/prakriti, vastu/resident matter of the site. Although with spiritual connotations, the outward expressions are very practical and aim at discovering the best suited layout and design for the place and purpose. The concept is often discussed as "genius loci" or the spirit of the site in Western discourse on architecture studies, just as the grids are, without explaining correlations with or acknowledging the antecedence of Vaastupurusha mandala and ancient Indian texts.

Site Plan and Layout

Vaastupurusha Mandala is a concept as well as a formal guide for organizing spaces. Spatially, the Mandala is contemplated and crafted as a spatial organization as well as structural design parti, with sub-grids of wide-ranging shapes and sizes to serve multiple purposes. Attribution of names of Devas/angels to each of these grid components expresses a religious connotation to the mandala; however most of these names are references to physical, geographical, and climatic forces. For example, in Mahapitha mandala, maruta refers to wind, shosha to dryness, mukhya to main, leading or facing, soma to moon, coolness or elixir, bhudhara to earth-bearing or mountain, aditi to mother of Sun or inexhaustible abundance; Rudra refers to Lord Shiva but also to heightened anger, Isha to Supreme, fertile or best, jayanta to victorious, apavatsa to one lacking offspring, aryaka to respectable men, aditya to the sun, bhrisa to severe or harsh, krishanu to an atom or part of Lord Krishna but also to darker soil, savitra to sun-related or fearing sun, vitaha to vanishing or useless, vivasvat to brilliantly shining with light, yama to God of death, bhringa to bee-king, pitri to father or ancestor related, Indra relates to the king of divine angels, Sugriva to neck or corridor or goose, conch and the like, mitraka to friendly, varuna to the angel of water, Brahma to the architect of the cosmos or genesis of cosmic construct. The position of the purusha as transposed on the Mandala grid changes with seasons from front to side and back. The relative proportions of Deva or physical, climatic aspect to specific grid component, with consideration of relative position to other grid or mandala components and the level of energy activation in each of those sub-grids, are emphasized under this concept. An 81 square mandala can be drawn in 1 square inch of space or 100 acres and such. When translated to English, most of the site plan mandala names are references to mathematics and geometry; the names and forms of these Mandalas corresponding with are shared in Appendix I. Appendix II presents illustrations of some of these site planning and town layout mandalas, as initially presented in Acharya's 1927a, b, 1934a, b, c, and 1946 publication series on Hindu architecture to give an idea of the shape and spatial organization potential of these mandalas.

A typical layout of the temple with reference to the town is illustrated in the diagram below (Fig. 2.2).

The temple precincts, Prakara and Parisar, of many pilgrimage sites, which have existed for centuries have been re-designed over time, to accommodate the nature and needs of evolving pilgrim. The precincts gradually acquiesce supplemental forms, patterns, amenities, and physical structures more suited to needs of

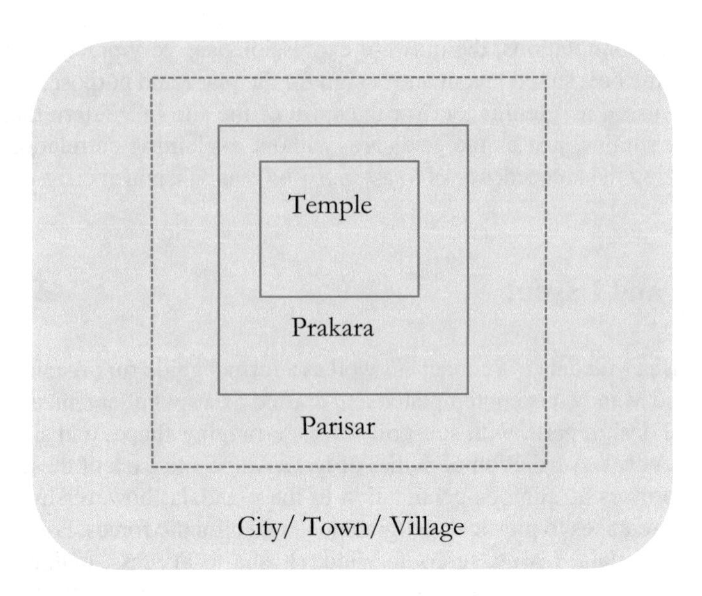

Fig. 2.2 Diagram of temple prakara and parisar

present-day stakeholders, the contemporary pilgrim and current temple administration. The physical design changes inadvertently have a consequence for landscape, urbanism, and ecology of the place; however, the discussion is missing in strength from the contemporary discourse.

Prakara is referred to as "courtyard" in Acharya, Vol 1, pgs 280, 298, and 325, as enclosed courtyard on pgs 412 and 634, as "an enclosure" on pg 395 while on the same page explaining the prakara as "a compound wall" thus alluding to "precinct" and as "city perimeter wall" on pg 391, and then to mean, "a number of" on pgs 15 and 129, "types" on pgs 74, 250, 761, and 766, parapets on pgs 174 and 198 quoting from Kautilya Arthashastra, possibly interpreted as "outdoor courtyard" on pgs 1 and 390 and ambiguously on pgs 389, 390, 411, 460, 482, 495, and 757. Francis D.K. Ching (1995), in his "A Visual Dictionary of Architecture" (253), also defines the "prakaram" as the temple compound around the sanctum.

Given the differential usage of prakara in current research, such as "outdoor compound" by Ching (1995) and Acharya (1927a, Vol 1) and "indoor courtyard" by Bharne and Krusche (2012) and Acharya (1927a, Vol 1), it is critical to clarify the usage of term "prakara" through this book. The term "prakara" would be used in this narrative to mean the outdoor temple compound markedly so. Parisar is the Hindi word used to indicate precinct or immediate surroundings. This area for religious sites is distinctly identified through the spread of formal or informal economic activities related to the temple; these are mainly concerning the prayer paraphernalia, flowers, prasad/sweets, coconuts, incense sticks, joss stick, kumkum the red vermillion, and pooja plate (consisting of all prayer items) and religious books, souvenirs, local arts and crafts merchandise, and food/drinks. The spread of these activities differs from one site to another depending on the religious significance and number of tourists per day and could be a range of one eighth of a mile to half a mile or approximately 200 meters to 1 kilometer. Hence, the term "parisar" would be used to imply immediate surroundings beyond the marked temple compound, for the spread of 10 minutes (200 meters or approximately 1/8th of a mile) to 40 minutes (1 kilometer or about 1/2 a mile) walking distance beyond the temple. The area beyond the parisar will be referred to as a town, as a representation of urbanized settlements, whether at the village, town, or city scale.

Site Suitability

Interpretations regarding auspiciousness in current Vaastu practice impress that the shape and slope of land plots are all critical considerations before initiating the purchase, design, or development of the site. The claims on auspiciousness need to be reviewed underscored by the understanding that these decisions are impacted by mathematical calculations of plot shape, size, and location within the broader context. While some rules are presented as cardinal rules, many are mutable as they are viewed against the numerological calculations and astrological constellation, planetary configurations associated with the owner or client of the property.

Shape and Slope

The plot shapes considered auspicious or beneficial for the inhabitant are all balanced, symmetrical shapes of the square, circle, rectangle, and complex expanding square grid, except the square or rectangular plot with an extension to the northeast (Sharma 2008). The shapes considered inauspicious or harmful for residents are a mix of symmetrical or asymmetrical oval, triangular, semicircular, spheroids, and polygons. These nominations seem to correspond with the climatic conditions and aspects of sun, shade, wind, and light; however, they are best tested through a series of scientifically conducted research (Table 2.1).

Table 2.1 Plot shape suitability (Sharma 2008)

Auspicious	**Lack of agreement in literature in terms of auspiciousness**
Square: Vargaakaar	Trapezoid with narrow front: Kaakmukhi
Rectangle: Ayaatakaar	Trapezoid with narrow front: Gomukhi
Equi-expanding square grid: Bhadarsan	Trapezoid with broad front: Chaajmukhi
Trapezoid with northeast extension	
Selectively auspicious	
Circular: Vruttakaar	
Hexagonal: Shatkoniya	
Octagonal: Ashtakoniya	
Inauspicious	
Triangular: Trikonakaar	Flattened hexagon: Mridungakar
Oval: Andakaar	Dumb-bell: Damroo
Semicircular: Gumbadakaar	Irregular pentagonal: Shakatakaar
Crescent shape: Ardhachandrakar	Circular-flask shaped: Kumbhakar
L-shaped	Trapezoid with south-east extension
Shapeless	Inverted t-shaped

Similarly, deductions regarding the slope as followed in the popular practice of Vaastu are annotated in (Table 2.2) below. Since these are also conjecturally climatically responsive but subject to verification through systematic scientific study, the annotation here is for reference purposes and to acquire an overview of variable sand aspects considered in the commonplace practice of Vaastu (purushamandala) Shastra, often abbreviated as Vaastu Shastra or Vaastu.

Table 2.2 Slope suitability (Abstracted from: Sharma 2008)

Name	Slope	Implications
		Auspicious
Gaja prustha/elephant back	Higher at S, W, SW, NW	Good for wealth, well-being, and longevity
Koorrma prustha/ turtle back	Higher at the center sloping down outward	Good for enthusiasm, wealth, and well-being
		Inauspicious
Daitya prustha/devil back	Higher at N, NE, sloping down to W, SW	Loss of health, wealth, progeny
Naaga prustha/snake back	High along N to S and elongated toward NW	Loss of health, wealth, progeny

Urbanism

Temples have been considered integral to the towns at least, since c300 BCE (Mookerji 1960). Arthashastra, also referred to as Kautilyashastra, lays out guidelines for organizing towns, and these principles as laid out by Kautilya Chanakya under

Chandragupta Maurya's reign were verified by scholars such as Mookerji (Mookerji quoting Arthashastra, 1960, 130), through works of Greek diplomats and historians' accounts of India and Indian texts, mainly Megasthenes. The towns were planned to have a balanced mix of public works and social institutions "aratnas (resthouses), prapa (tanks), sattras (alms-houses), pavyasthanas (holy spots), chaityas (trees for worship), and deva-grihas (temples), besides housing for residents. Additionally, there is an emphasis on the construction of structures solely for the decoration of the village. The limits of the towns were marked by natural boundaries such as river, hill, forest, shrubs (grihati), valley (dart), embankments (tatbandha), and trees like the silk cotton tree (shalmali) from the Salmalia malabarica family (Mookerji 1960).

Pali texts and Jatakas (Mookerji quoting Pali texts and Jatakas, 1960, 198) also underscored the town layout principles. They ordained that the village (grama-kshetra) was to be bordered with the arable land, grazing grounds, or pastures beyond the arable fields, for cattle, then the groves, just like the Veiuvana at Rajagriha, the Anjanavana at Saketa, or the Jetavana at Sravasti. Fences and traps (for rodents) were installed at the border of arable land with the pastures to prevent pests and cattle. After the pastures, a belt of uncleared jungles or forests was planned to be used as a resource for the firewood for the village and as hunting grounds, and this could be a combination of forested or unforested wastelands (akrishihya bhumi) lying beyond the village. The forest belt most proximate to the village could contain woodland retreats or meditation forests—tapovana, for Brahmanas for their study of the Veda and performance of soma sacrifices and penance (Mookerji quoting Arthashastra, 1960, 201). Lastly, the village is to be bordered by either an upasalam, a palisade of stone or wood posts, or a wall or stockade with gates or gramadvara (Fig. 2.3).

Manasara defines eight kinds of shapes for villages, Dandaka, Chaturmukha, Nandyavarta, Prastara, Padmaka, Sarvatobhadra, Svastika, and Karmuka (Acharya 1927a, b, 63), each of these has a variety of expressions in Mandalas. This typology of town plans was devised based on the various sizes and scales of site plan mandalas explained in the earlier section. The shapes and Mandalas for the towns, which are known as nagara in Hindi and Sanskrit, and can translate to contemporary cities, are mentioned with reference to the seat and residence of the King or the ruling, governing body. They are known as Rajadhaniya nagar, Kevala nagar, Pura, Nagari, Kheta, Kharvata, Kunjaka, and Pattana (Acharya 1927a, b, 95). The villages and towns had a specifier of pura, nagara, and kostha in correspondence with the number of inhabitants. A Kostha would indicate a settlement of 100 households while a Nagar would have 50 households. This is akin to the designation of neighborhood, town, and city only if they corresponded with the number of households or inhabitants and this was standard for all settlements across the country.

Here is a brief overview of site and town planning Mandalas, with an exclusive focus on the spatial organization and pattern. Reading the description and the graphic, through the eyes of an architect, the pattern, and the embedded spatial organization seems to have been iteratively developed, with the pattern being the starting point.

Dandaka: Predominantly one quadrangular grid, with a simple form and structure, vertically and horizontally subdivided by up to five major carriageways mostly having a uniform partitioning and single axial emphasis, if any.

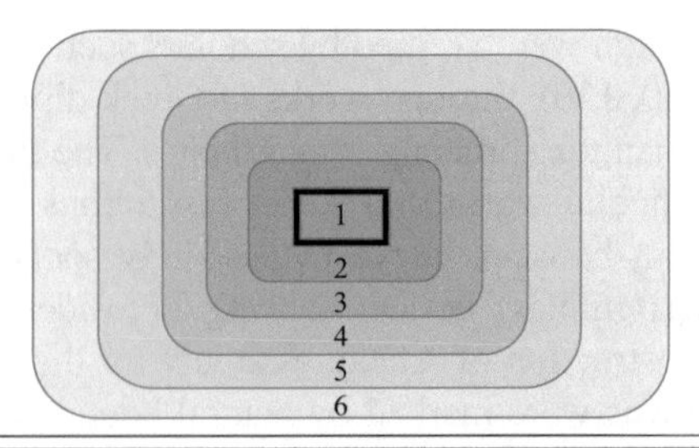

Key: 1: Village, 2: arable land, with fence/snare wall, 3:pastures, 4, 5, 6, Belts of forests, 4: Tapovana- Meditation forest or Brahma-Somaranya – the forest for offering sacrifices, 5: Vihara: King's hunting forest, 6: Resource producing forests, named based on the product: distinguished by their products such as Daru (timber), Venu (bamboo), Valli (cane), Valka (bark), Rajju (fibres for robe-works) ; Patra (material for writing such as palm-leaves or bark of birch, tala-bhurja-patra); Pushpa(flowers for dyeing such as Kimsuka, Kusumba, Kunkuma); Aushadha (medicinal herbs), Visha (poisons) [II. ry], firewood, and fodder (Kastha-yavasa), and Vajravana: forests for elephants -a critical tool for war while also providing timber for town-building and fortification.

Fig. 2.3 Diagram of village layout within the broader landscape, circa 323 BCE, during Chandragupta Maurya and Chanakya reign

Chaturmukha: Temple at the center, three predominant rectangular grids, inside-out, the center most grid is un-penetrated, the middle grid is punctuated by a diamond shape roadway, the external grid is subdivided into three vertical grids and three gateways as well as carriageways, the external two grids have an east-west axial carriageway running through them.

Nandyavarta: Has a temple in the center, like Charturmukha, except that predominantly there are two quadrangular grids, vertical grid demarcations are softer, and there is no diamond shape punctuation in the middle tier grid. Nandyavarta quadrangles can also be laid out in the form of 90 degrees rotating grid, akin to Svastika.

Prastara: with relative reference to Chaturmukha, there is no dense center to the grid, the central grid is softer as surrounded by smaller carriageways, the external-most grid has three vertical grids and three gateways as well as carriageways, and all three grids have an east-west axial carriageway running through them. The layout seems more densely built as more spaces between the grids are attributed with a use/inhabitation.

Padmaka: A large quadrangle, punctured in the center by another quadrangle with a center marked by another much quadrangle similar to the dimensions of horizontal and vertical axial carriageways. The large quadrangle is divided into four subdivisions with even further partitions within each of those four quadrangles and is surrounded by a poly-shape quadrangle with circular corners.

Sarvatobhadra: Like Padmaka with, with a softer marking of the center, and finer variations within the predominant quadrangle and further sub-partitions within each of the four quadrangles, except that the surrounding grid is not polygonal but straight-edge and can read like a part of external quadrangle.

Svastika: A quadrangle with a marked center with two predominant grids, both intercepted by an axial carriageway running north-south and east-west, the internal quadrangular is 90 degrees rotating grid.

Graphic illustrations of key Mandala are redrawn based on Acharya (1927a, b) and Sharma (2008) and presented as a referential visual cue on the spatial organizational blueprint. The mandalas are diagrammatic retaining only the basic information of pattern projecting the spatial organization.

Dandaka (Sharma 2008)

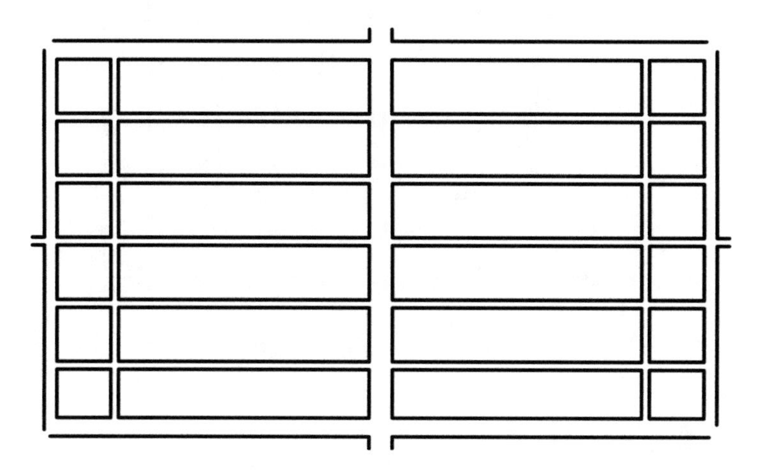

Chaturmukha- village layout (Acharya 1927a, b)

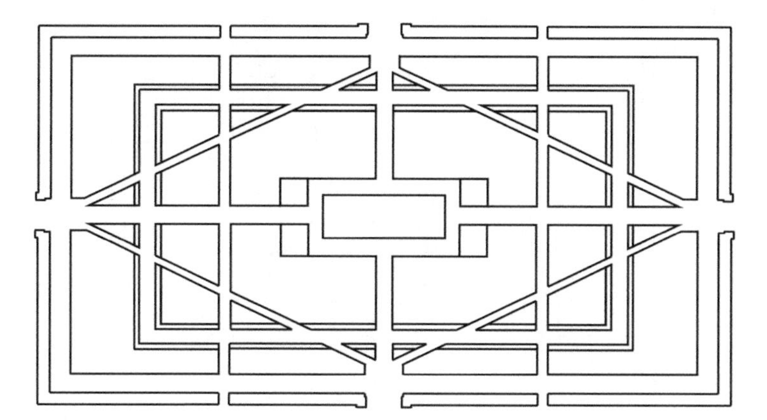

Nandyavrut (Sharma 2008)

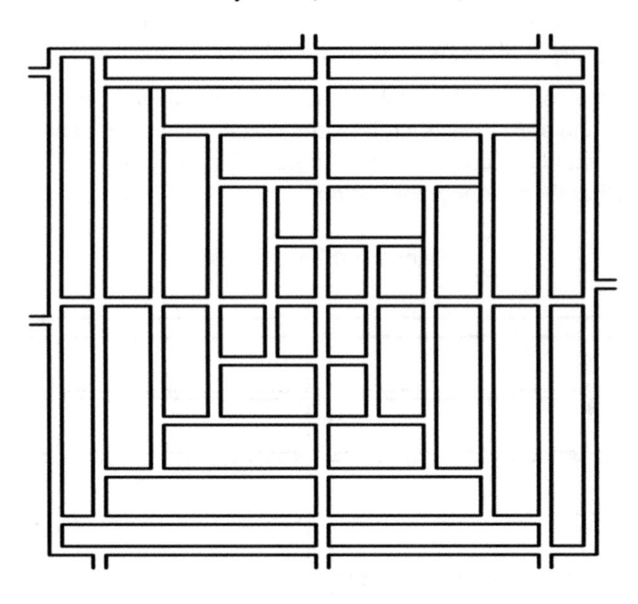

Prastara (Sharma 2008)

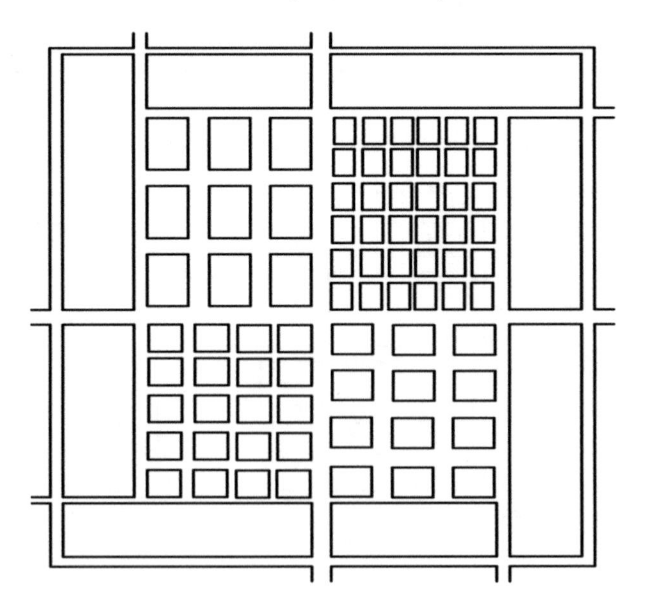

Padmaka (Sharma 2008)

Sarvatobhadra, radial (Sharma 2008)

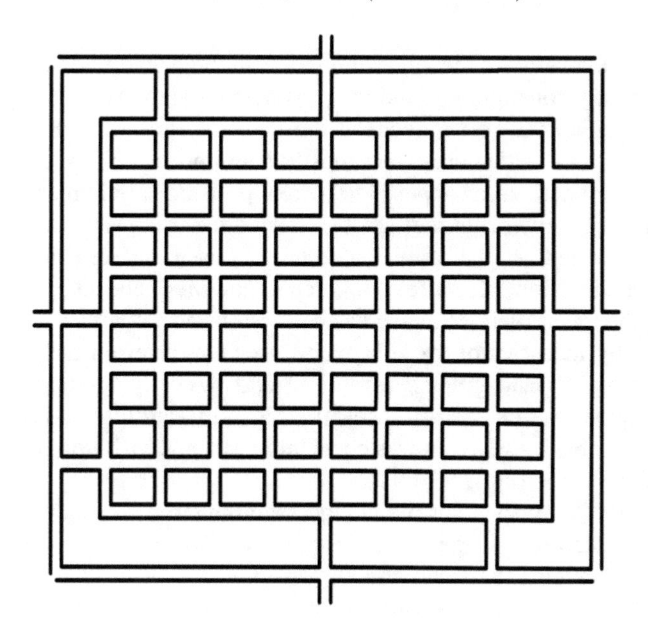

Svastika (Sharma 2008)

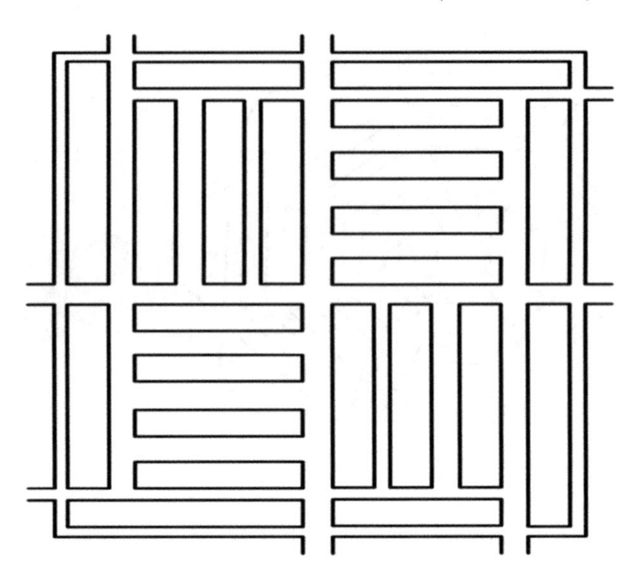

The distinction between the village and town in India was predominantly associated with primary "operational functions" associated with them. The geographical scale of the settlement also influenced the determination of attributed function. Initial pre-industrial settlements were mostly being planned for specific economic and operational functions, as opposed to current post-industrial times where there is a high supply of settlements that are best retrofitted to newer purposes. Villages were often planned as the rural republics, focused on agricultural purposes (Mookerji, quoting from Metcalfe's 1832 report and Megasthenes' accounts, 1960). In contrast, the towns were conceived as centers of trade, commerce, and administration. The physical scale of the village and town was determined in ancient times mainly in terms of land granted for tilling and basically for taxation purposes (Mookerji, quoting Manusmriti, Metcalfe, Kautilyashastra, and Megasthenes, 1960). Land tillable by six oxen was considered as equal to 10 villages and a 1000 villages made a town (Ibid).

Acharya (1927a, b) specifies the village extant in terms of a habitat for a 100 to 500 families, primarily engaged in farming, with a maximum spread of 2250 yards and towns as ranging between the dimensions of 100 × 200 × 4 cubits to 7200 × 14,400 × 4 cubits (Acharya 1927a, b, 55, 284). To get a sense of scale in current times, 2250 yards is roughly 22.5 times the length of an American football field. Once the purpose was determined, the shape, type, and proximity to the surrounding landscape also influenced the decision-making on which mandala could be chosen for the site from the broader coterie. For example, Karmuka is a semicircular layout pattern for waterside locations while Kharvata is a circular fortified pattern for plains or mountainous surroundings. Kubjaka and Khetaka are layouts for subdivisions within the villages and towns.

Karmuka—village layout (Acharya 1927a, b)

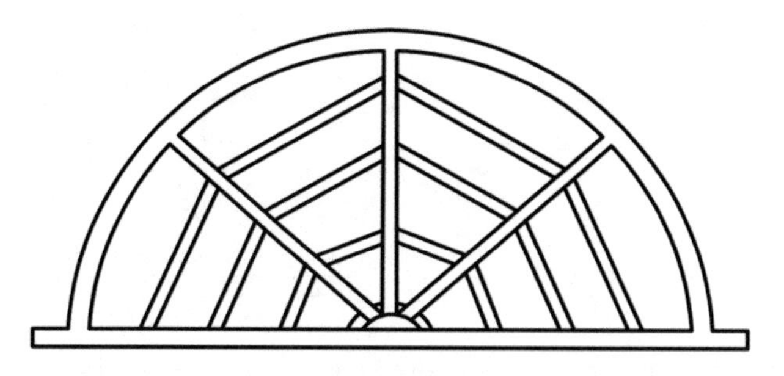

Kharvata—town layout (Acharya 1927a, b)

Khetaka—waterfront town layout (Acharya 1927a, b)

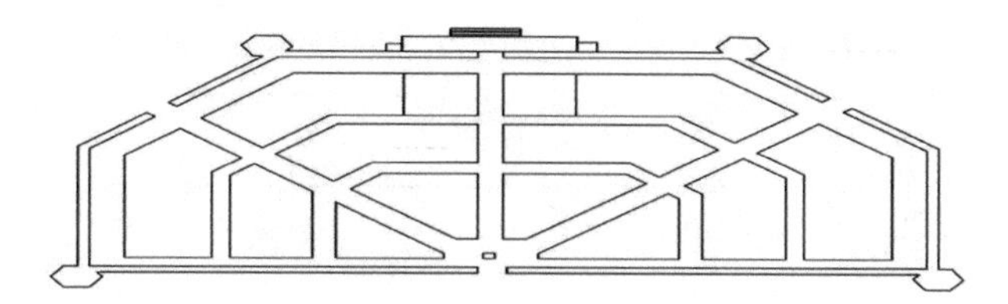

Kubjaka—town layout (Acharya 1927a, b)

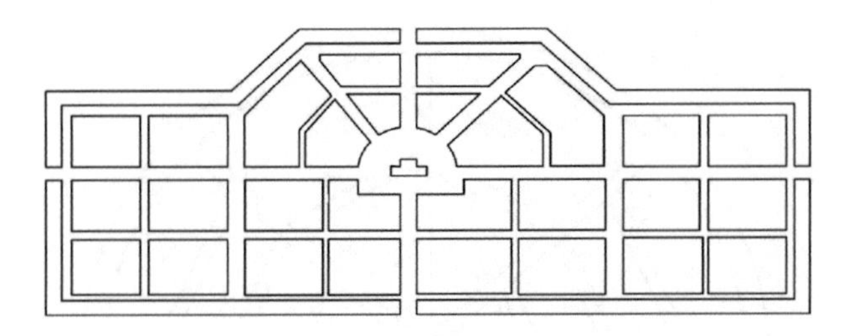

A triangular layout pattern of Paramasayika mandala is also mentioned in texts related to Vaastu texts. This reminds us that the organization and divination of mandala patterns in Vedic civilization fundamentally started with organization pattern for "vedi" or an altar to offer prayers through *yagya* or *havan*, through the lighting of woodpiles on top of the altar.

Paramasayika, site layout plan (Acharya 1934a, b, c, Series 5, sheet no XII)

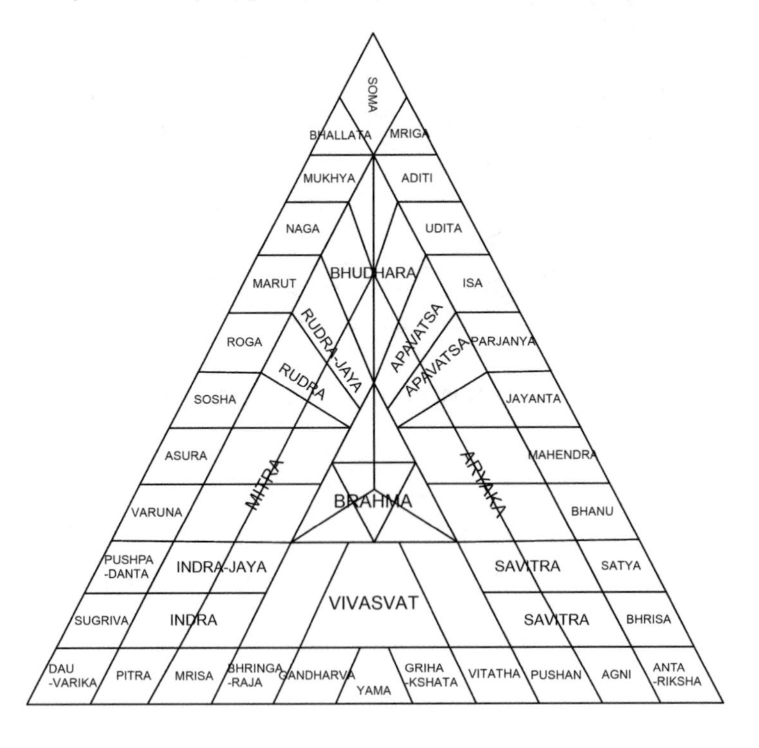

Landscape

The villages and towns were not only planned for economic self-sufficiency but also be endowed with all public conveniences and landscape to accommodate visitors. The landscape should include shade-giving and fruit-bearing trees, gardens of medicinal plants, rivers, lakes, and water tanks (Mookerji 1960, 74). A variation of landscaping is recommended for the village residents. The guidance includes flower gardens, horticulture gardens, orchards (pushpa-phalavada), clusters of trees such as lotus or bamboo, and farming fields (Mookerji 1960, 124). The guidance of farming within the village includes flower plantations (phala vana); fruit orchards (phala-vada) such as of bananas, sugarcane, and the like; and mulavapa or fields for growing roots like ginger, turmeric, and similar spices (ardrakaharidradi). Thus, the ancient texts' advisory on growing grains, flowers, fruits, vegetables, spices, sugarcane, and bananas in the village (Mookerji 1960, 197) not only aimed at self-sufficiency of the village but also as a food, flowers, and water, supporting towns.

Soil Suitability

The health of vegetation based on soil and landscape context and elemental composition (panch-mahabhoot) of soil (Srikanth et al. 2015) is:

1. Prithvi pebbly dark blue or black rich vegetation
2. Jala unctuous, cool white grass
3. Agni stony multicolor small sized trees
4. Vayu rough gray small trees
5. Akasha soft no color trees of no value

Soil Type

This classification from Sharma does not completely reconcile with the soil classification based on color and taste as presented by Srikanth et al. (2015, 397) shared below, for red and off-white color soils (Table 2.3):

- Black (Asita) sweet most fertile
- Off-white (Vipandu) sour less fertile
- Blue (Syamala) salty relatively lesser infertile
- Red (Lohita) bitter in-fertile
- Yellow (pita) astringent relatively infertile

It is also stated that lands that are "beaten by foam" (phenaghuta), i.e., those on the banks of rivers or marshy lands, are suitable for growing water-loving fruits such as pumpkin and gourd. The flooded lands (parivahantah) are suitable for sugarcane (ikshu), pepper (pippali), and grapes (mridriksh). Lands closer to

Table 2.3 Soil types and connotations

Name		Auspicious	Inauspicious
	Color	White, golden, red, black	Yellow, off-white, blue, red (?)
	Odor	Ghee like: for Brahman—education, religion, philosophy Blood like: for Kshatriya—soldiers Grain like: for Vaishya—business Wine like: for Shoodra—scavenging	
	Taste	Sweetish—for Brahman Savory—for Kshatriya Sour—for Vaishya Bitter—for Shoodra	

water-wells and tanks (kupa-paryantah) are good for growing vegetables (kwathita) and roots (mula). Lands in the vicinity of canals, lakes, or tanks (haraniparyantah) are good for fodder (harana). Lands between cultivated plots may be utilized to grow cloves, medicinal herbs, and fragrant plants. Medicinal plants arc to be grown on different kinds of lands, marshy and dry, as required.

Water

Water is central to ritualistic practice of the Hindu faith and has been accorded a critical place in design planning principles as well. Construction of water features is prescribed in various ancient texts such as Matsyapuran and Vaasturaj Vallabh (Sharma 2008, 187; Shukla 1967, 26–27). The recommendation is for construction of at least four vapis or ponds; ten Kunnye, koopa, or wells; four kundas, perhaps basins; and six tadagas or taalaab, loosely translated to reservoirs or lakes at the center or periphery of the towns; these include water supply considerations for birds, cattle, and humans. In residential context, there are specific instructions on locating water feature or water source at only northern and northeastern sides as per Varahamihira text for auspicious outcomes, whereas Vishwakarma text adds West direction to this. The suggested shape of the wells (also referred to as Koopas) is circular.

In terms of shapes and size typology, the Koopas are said to be of ten kinds each with a specific stylistic name with correspondingly varied dimensions but are all circular in form (Sharma 2008; Shukla 1967, 26–27). The dimensions of Koopas range from 4 hasta to 12 hasta with structures less than 4 hasta being called Koopikas. Vapis or lakes are prescribed to be of four styles with ek-vaktra (vaktra = aperture/ mouth) and tri-kuta (kuta = peaks or base), dvi-vaktra and sat-kuta, tri-vaktra and nava-kuta, and chatur-vaktra and dvadas kuta. Kundas (Shukla 1967, 26–27) are said to be a religious structure with offsets or insets in construction to place a deity. The six types of tadagas as described in Shukla are the following: Sara, half moon shape; Mahasara, circular; Bhadraka, square; Subhadra, with excessive bhadras;

Parigha, Bakaikasthala (Parigha = surrounding/estimation/enumeration; Baka = crane or *Sesbania grandiflora*, Bot. tree; Sthala = place/habitat), and Yugma parigha, Bakas (crane abounding on both the banks).

The recommendation on a precise number of these water features in a particular town cannot be consensually concluded (check Shukla 1961) based on these above descriptions; however, reading the content in the context of guidelines on town planning, it is safe to conclude that the number was derived at through considerations of the size of the town and inhabiting population. However, certain water features could be considered a factor in siting the temple. Based on the sloka quoted from "Patala, 19" by Rao (1997, 132) and presented below states an instruction on locating the temple near "tadaga" if not beside the sea, mountains, forest, or grove. The English transliteration of the sloka is "Nadhyadrisamudraparshve anyatra tatakaaraamasameepe veeviktavanadeshe vaal (1, 9, 3)".

The instruction on pleasure ponds and lotus lakes is mostly concerning palatial gardens rather than temple context. There are references to water systems through the Vapis, Panagrha or watersheds, and other animal species such as the sheds for horses, elephants, and cows (Shukla 1967, 6, 7, 12). Watershed for both drinking and other uses as Panagrha (Shukla 1967, 6, 7, 12).

In addition to the unique water feature, there has been reference to the distribution of water from these features through water sluices and canals to all residents and for irrigation (Mookerji, quoting Kautilyashastr and Chanakya, 1960), indicating attention to the networking of these features for democratic distribution of water resource, thus ensuring the verdant health of village or town as a whole.

Vegetation

Reviewing the orientation of Mayamatam, one of the key ancient treatises on town planning and temple design as presented by Sastri (1919, 1, 2), it becomes clear that this particular text did not focus on details of vegetation around the temple but on built constructs. Popular religious, cultural practices, and contemporary texts prescribing vegetation specifics, will thus be drawn from for this section.

A multitude of Hindu rituals is based on tree and river worship. This is an expression of the general attitude of natural resources conservation as ingrained in the Hindu culture, independent of co-location with the temples. The cultural practices have anecdotal repercussions woven into the narrative, which could imply that religious rituals were used to instill nature conservation values in people. For example, it is said that a man has ancestors suffering in hell; fourteen out of them are sure to be redeemed if this person plants five mango trees either in the garden or on the roadside or as mentioned in *Tharumahima.* The values of conserving vegetation and water are ingrained in religious, cultural narratives.

Another example is the saying that ten wells are equivalent to one pond, ten ponds to one lake, ten lakes to one son, and ten sons to one tree, thus extolling the virtue of planting a tree and constructing ponds and lakes, but also emphasizes on tree planting as an extremely significant act (Srikanth et al. 2015). Rigveda since

5000 BCE has offered prescribed guidance for people to be kind and caring to all vegetation, especially those enriched with medicinal plants (Srikanth et al. 2015, 390; Narayanan 2001). Vriksha Ayurveda, a component of Agnipurana, attributed to Salihotra around 400 BCE, a dedicated text on ancient Indian agricultural science, signifies the conservation of plants and equates ten human lives to a single tree. Mookerji (1960, 109) notes that King Asoka attached immense value to medicinal plants for the health and well-being of towns (including humans and animals). As recorded on his rock edict, he advocated for the cultivation and conservation of medicinal plants in the State Botanical Gardens.

Ideally, a village plan should have belts of (i) pastures (vivita) for the grazing of its cattle; (2) sylvan retreats for religious study, practices, and meditation; (3) a reserved forest for royal hunting to be stocked "with tamed (danta) animals like deer and elephant, and wild animals like tiger but with their teeth and claws cut off"; (4) forests as habitat of all animals (sarvatithi-mrigammri ja-vanam); (5) commerce-oriented forests for growing different kinds of produce such as timber forests, bamboo forests, or forests of bark-producing trees; (6) factories using forest products as a resource (dravya-vana-karmantan); (7) colonies of foresters; and (8) forests for rearing of elephants beyond human habitation" (Mookerji 1960, 124, 95). Commerce-oriented forest was further refined in subcategories, depending on type of timber, such as Daru-vana, for construction timber such as Sola, dividapa, etc.; Venu-vana, bamboos of different kinds; Valli-vana, creepers (vel, vellika-vana), different types of fibers such as hemp (hajju-bharjida) for ropery; patra or leaves for writing, such as palm leaves; flowers as materials for dyeing, "such as kimiuka, kusumbha, and kupkuma"; and aushadha-vana, medicinal plants yielding herbs, roots, and fruits (kanda-mula-phala) used as medicines.

Attention to plants and gardens around palaces is noteworthy as they share the edge with ordinary folk residences. Folly-like structures, mazes, secret and subterranean passages, hollow pillars, hidden staircases, and collapsible floors were designed in proximity to palaces of residences of noblemen for security purposes as a distraction, and plants for pavilions were selected for poisonous nor an attraction for poisonous animals. For example, the use of trees that are avoided by snakes and the inclusion of trees that host parrots because they cry out an alarm signal at the sight of snakes. The creeper pavilions are known as lata griha. Wooden hill- darugiri, the vapis, and the well-laid flower lines are called puspa vithis, and the flower pavilions are Puspa vesma, together with the machine room yantra karamanta (Shukla 1967, 6, 7, 12).

The planting typology prescribed by Kautilya resonates with Manu's recommendations on vanaspati, fruit-producing without evident flowers and oushadhis—medicinal plants (Srikanth et al. quoting Manu's plant taxonomy, 2015). It also describes the plants based on their physical characteristics and canopy cover, Vrikshas, trees producing either flowers or fruits; Gulmas, shrubs that spread several branches a little above the ground such as *Nerium*; Gucchas, bushy shrubs such as jasmine; Truna, grasses; Pratanas, creepers spreading on the ground; and Vallis, creepers that twine and climb around other trees for support; these typological recommendations for villages applied to planned settlements bordered by the forest belts. Vrikshayurveda mentions specific tree species ("Peral" in the east, "Athi" in the south, "Arayal" in the west, and "Eithi" in the north) to be planted around the residence but not such that the branches and leaves reach the roof and windows (Srikanth et al. 2015).

Recreational Gardens, Parks, and Amusement

There is a prescription for installations and structures in the landscape around the built constructs but mainly for palaces. Dharagrha or shower bower/water fountains are recommended to be sited close to big reservoirs in a beautiful surrounding in the palatial gardens (Shukla 1967, as quoted from Samrangana Sutradhara, 45, 47). Jalakreeda refers to amusement structures in these pleasure gardens, with peacocks, swans, monkeys, snakes, Kinnars, and vriksha, or the imagery of the aforementioned flora and fauna. Jalamagna is an underwater chamber for the King at the bottom of a deep-water reservoir with a continuous flow of water above to keep the chamber cool and secret; the reservoir should have mechanical lotuses, fishes, birds. The Nandyavarta maze-like yantra installation is submerged in water for pleasure-play. The merry-go-round mechanism style of equipment was installed on the ground.

Ecology

Ecology essentially means the study of inter-relationships of inter-connected systems. This Western conceptualization of ecology has been attributed to Ernst Haeckel (Seidler and Bawa 2016; Egerton 2013; Balgooyen 1973; Friederichs 1958). Haeckel's definition of ecology, as presented by Klauuw and Meyer, 1936, is shared as an English translation (courtesy Google translate) below,

> *...ecology is the science of the relationships of the organism to the surrounding outside world, the physiology of the interrelationships of the organisms to the outside world and with each other, thus also the science of the interrelationships of the organisms with each other. In terms of ecology, he broadly includes all "conditions of existence"; they force the forms of the organisms to adapt to them. What is meant are inorganic conditions of existence, such as physical and chemical properties of the place of residence, climate, inorganic food, water quality, soil, etc. Under the organic conditions of existence, he counts all relationships of the organism to all other organisms with which he comes into contact.* (Klaauw and Meyer, presentation of Haeckel's 1866 definition of Oecologie, 1936, 147)

Understanding the idea of ecology from ancient scriptures requires an interpretive reading of the text and an implicit understanding of the Hindu religion and philosophy. The interconnectedness of organisms in terms of mutual impacts is tied to religious rituals and as Jain (2011, 2019) reminds us, is most often passed over as dharma or duty (inadvertently responding to Narayanan's (2001) call for making these inter-connections, and impacts more explicit for the masses to reign then back to an ecologically responsible way of life. The idea of honoring ecological forces and planning with respect to those is central to the concept of "Vaastupurusha mandala." The idea of "purusha"—the cosmic force—and "prakriti," the physical or formal, has been intrinsic to the Vedic philosophy since around 7500 BCE; some schools of thoughts attach converse meanings to Purusha and Prakriti, which is best left to Darshanshastris or philosophers, but there is an agreement on the presence of the two. Noble (1981) acknowledges the same although discussing temples in Southern India, but the time when the distinction between the divine God and nature (prakriti) is blurred almost enough to disappear, thus doing away with the dichotomy. Vedic slokas offering venerations to natures are dedicated to the panch-mahabhoot: Prithvi/earth, Agni/fire, water/Aaapah; air/Vayu, and akasha/space present an understanding of the five fundamental elements of nature. The element of dhatu was integrated with the approach to foundational elements, during the Bronze Age, with evidenced glory though the Indus Valley civilization and Ayurveda approximately attributed to 1500 BCE that uses the concept of dhatu—matter for discussing the composition of the human body.

The panch-mahabhoot is considered the building block of the cosmos. The concept is where everything is composed of panch-mahabhoot and is thus interconnected and mutually impacting presents a Vedic view on the ecosystem. Different combinations of the panch-mahabhoot create various forms and systems with varied characteristics when permutated and combined differently. An imbalance of any composite element creates a Dosha or what we may call an imbalance or contamination. The contemporary field of ecology is constantly testing the extent, intensity, and effects of the connections between the purusha prakriti.

Vaastupurusha mandala is a tool to inform siting design and construction decisions to sustain or create a harmonious balance of the five elements—the panch-mahabhoot for the site and its surrounding context.

Chapter 3
Other Interpretive Frameworks

Nomenclature

Given that this discourse is presented in the English language, it is a natural step to first account for the common understanding of temple as in English language and Western culture. Encyclopaedia Britannica (2019) defines the temple as "an edifice constructed for religious worship." It notes the references to temples in Roman and Greek times. In Merriam-Webster dictionary (2019), temple is defined as "a building for religious practice." It then explains the term with examples of Jewish and Mormon temples. None alludes to origins or examples in the Indian context and Hindu culture. The omission does not alter the fact about the flourishing Hindu temple construction as indicated by the temple ruins and structures, since 1500 BCE (TOI 2021), and design treatise since the key ancient Indian texts, with even the most recent of the texts Manasara attributed to 550 BCE (Acharya 1927a, b, 1934a, b, c).

In Hindi and Sanskrit, as indicated by the Sanskrit dictionary (2019), temples or places of worship have been referred to with a range of names such as devālayaḥ, īśvarasadan, devakulam, devagṛuham, devabhavanam, deva-vaas, devāvasathan, devāgram, devayatana, devatāgram, puṇyagṛuham, pūjāgṛuha, maṅgalagṛuha, and mahālayaḥ, to broadly mean "the abode or place of God," based in approximate transliterated meanings Deva = God or Godlike to imply heavenly, divine, and used for humans of high excellence, alaya = dwelling/place, isvara = God, kulam = family, punya = good act, gruha/sadan/avas = residence, maha = great, mangala = sacred, puja = prayer, ayatan = seat/place/shed for sacrifice/place of sacred fire. It is not clear as to what Vedic text used the word devalaya for the first time and furthermore, when was it taken over by the currently more popular form of "mandir."

Mandir shows up in Oxford dictionary which defines it as *A Hindu temple, with origin from Hindi and Sanskrit Mandira 'dwelling place, temple'* (Oxford Dictionary 2019). Mandir shows up with similar meaning, in Merriam-Webster as well, but not

A. Sharma, *Mandala Urbanism, Landscape, and Ecology*,
https://doi.org/10.1007/978-3-030-87285-4_3

Table 3.1 Nomenclature of temple from Hindu Dictionary of Architecture by Acharya (1927a)

Sanskrit terms used for temple	Meanings
Devayatana	A Temple. Shelter for the "Deva" (Acharya 1927a, 230)
Devalaya	A god's residence or dwelling (Acharya 1927a, 231, 232)
Mandira	A type of building, a hall, a room, a temple, a shrine. (Acharya 1927a, 413–414)
Deva-kula(-ika)	A chapel, a shrine, a temple, a statue shrine, a statue gallery. (Acharya 1927a, 229)
Deva-niketa-mandala	A group of temples (Acharya 1927a, 263)
Mandara	A type of building which is 30 cubits wide, has ten stories and turrets (Acharya 1927a, 413)
Puri	A temple, an adytum, a building, a town. (Acharya 1927a, 312)

in Encyclopaedia Britannica, but with no circular reference back to the definition of temple. The definition of temple shows up as similar to that of Merriam-Webster, with the exclusion of reference to Indo-European from the history of etymology part.

The Hindu Architecture Dictionary by Acharya (1927a) provides the Sanskrit terms for Hindu temple or place of worship; see Table 3.1 below. Analysis of the semantic meaning of these Sanskrit terms and cross-checking the meaning with the description of corresponding examples renders an understanding on typology of Hindu temples, beyond the architectural style.

Typology

Style and *type* are to architecture what taxonomy is to biology, a system of classification and order to understand the subject in terms of individual or cognate entity, meta- and sub-derivatives thereof. Style by definition allows for subjectivity at two levels for categorizing, either a product or a process; thus the focus is on unique characteristics with a focus on differentiation. On the other hand, type mainly applies to the unity of intrinsic characteristics.

According to Lorenzetti (2015), architecturally speaking, style and type are the main defining aspects of typology. Typology refers to the system of classification that adheres to the study of a multitude of items that relate to the subject of review, current findings of the investigation and observation based on repetitive patterns, and lastly, ordering objects of similar designs within a single typology (Lorenzetti 2015).

Typologies are a dominant feature in the Sanskrit discourse that survived the medieval and ancient eras. The canonical texts or manuscripts known as Vastuśāstra are not an exemption to this. The classification and naming of all the forms of settlement such as altars, temples, stables, houses, and gods' palaces were underscored by elaborate creative and strategic thinking (Tillotson 2005). The distinctive categories and varieties are integral to the manner in which Indian temple construction was traditionally regarded, as is apparent from the architectural texts. The formal typologies are the foundation of the temple design in both the variations of the given types and the assemblage of composite designs. While currently the general planning of the Indian temples as contextualized within the broader geographical as well as cultural context does not reflect on the typological ways of thoughts, it is the Prasad, the shrine, as well as the encompassing envelope and the personification of divinity which remains central to the ideology behind the temple (Tillotson 2005).

Acharya observes that the discussion on "style" was a later projection rather than the core concern of the Manasara (Acharya, Vol 2, 1927b, 180). He bases his observation on the fact that there is no elaborate discussion of these in Puranas but scattered mentions in Brihat Samhita, some Agamas, and relatively more detail in Manasara, Kamikagama, and Suprabhedgama. Hindu temples have broadly been categorized as Nagara, Dravida, and Vesara, where Nagara style had quadrangular shape, Dravida has a round shape, and Vesara has an octagonal or hexagonal shape. Here the reference to shape can be interpreted as a reference to the "shape of the architectural plan" of the temple, which may not reflect in the temple's elevation or cross-sectional view. The three styles represent the geopolitical Northern, Southern, and Eastern practices in temple design construction. The fourth style of Kalinga was found regarding the design of cars and chariots on an inscription on a column of Amritesvara temple however the inscription mentions Kalinga as a type of building.

Nagara and Dravida temples are primarily identified using both the Northern and the Southern temple architectural styles, respectively. The textual context of both Dravida and Nagara, which are present in Sastra texts, do not operate as all-inclusive stylistic approaches. Still, they are indicative of the general impulse for classifying temples based on their typological characteristics. The religious rituals that mainly lie at the core of the Hinduism religion, and the seemingly indefinable beliefs accompanying the Hindu divinity for some, bears a significant influence on the type of temple architecture designs, constructed both in the modern and ancient periods (Sinha 1996).

Nagara holds a unique temple style characterized by a shikhara, which is a building or a tower. In northern India, the typical Nagara temple is a small-sized and square-shaped sanctuary known as garbhagriha. It houses the main image and is preceded by adjoining halls connecting to the sanctum (Sinha 1996). The entrance of the sanctum is highly decorated with geometric ornaments and figures. The shikhara is placed right above the main sanctuary. One of the typical northern temple styles can be observed in the ancient Orissa temples like the Parasuramesvara temple constructed in the eighth century at Bhubaneswar which is a popular city for its intense temple building operations.

On the other hand, the southern Indian temple architectural style or Dravidian has a commanding gateway. The best example of the temple style is the Gangaikonda Cholapuram and Rajarajesvara temples. The construction style consists of a pyramidal structure. The style is mainly seen in the historic temples that adhered to the typology (Hardy 2001). The temple hosts a square chambered sanctuary, mandapa, and a rectangular court. The walls on the outside are held together by pillars (Sinha 1996). The tower on top of the sanctuary has gradually receding stories that occupy the pyramidal shape. Every story or floor level is described by the wall of the miniature shrine with rectangular-shaped roofs.

The temple is said to be of three types: sandhar, nirandhar, or meru prasad. There are detailed calculations regarding each of the architectural elements and the enveloped spatial construct; however it is not clear as to what philosophy, need, and purpose generated these precise mathematical calculations. Jain (1936) quotes Prasadamandana and accurate mathematical calculation for arriving at the geometry of the Shikhara form and variations. Reference to the building material of the Prasada/temple is mentioned as brick, stone, and timber (Jain, 1936, 120, 121 as below; Kramrisch, Vol 1- Part IV, 1946, 76). However, Sompura and Nathji (1965, 31) indicate that Mitti (Mud), Kastha (timber), Ishtika (brick), Shila (stone), Dhatu (metal), and Ratna (gemstones) are also materials to be considered and used per economic capacity. Types of timber deemed best or auspicious for the temple construction are also listed (Jain 1936, 121). There is an emphasis on the shuddhata/positive qualities of the client or devotee, acharya, sthapati, artisans, material, etc.

Hardy (2007) interprets the morphology of Hindu temples and unpacks mathematics of design in Hindu temples thus extending the inquiry of mathematics in ancient times such as by Williams and Ostwald (2015) but framed through an architectural perspective. Five types can be deduced from architectural morphological studies by Hardy (2007, 2013): morphological or descriptive, projectile point, chronological, functional, and stylistic typology (Hardy 2013; Gandotra 2011). Morphological typology is focused on the physical attributes and external characteristics of an object. An example of this type of typology entails the classification of artifacts on the ground of height, color, weight, material, and other features that the person might choose (Sinha 2000).

A projectile point is a form of typology whereby the archeologist narrows down the classification by sorting artifacts based on descriptive features such as color, weight, or as central to the research goals (Carmean 2009). For example, the classification is done based on shape in most cases. Repetition is a feature that unavoidably describes the overall stylistic evolution of temples in India. The Dravida temple created a rather complex wall system. Repetitive projections are evident with notable pillars, making the structures distinguishable (Hardy 2009). While the details of constructing the walls might appear to be the same, the architectural choices continue to separate them. For instance, the northern temples rely on horseshoe-like windows shapes, which can also be seen in temples built later in their moldings or decorations. The classification, however, should not be confused as focused on the morphological appearance (Jain 2005). Pottery is deemed a good example of how the typology works, given that the objects also offer adequate details about the chronological evolution of the artist and artistry.

Chronological typology refers to the sequential organization of artifacts based on their distinctive forms (Vandyck and Bertels 2017; Hardy 2013; Moudon 1994).

The process involves collecting data or related dates that are useful in establishing their actual positions in terms of time. The classification helps in reflecting on events that took place within the region. Chronological typology normally provides suggestions on when certain people or events took place concerning a specific time. On the other hand, functional typology entails the organization of artifacts based on their value or the purpose they serve instead of focusing on their appearance or the chronological sequences they own (Vandyck and Bertels 2017). In some instances, the objects might not be removed based on their functional objectives, and restoring them might prove to be a challenging process.

In the earlier Puranic Age, people involved in building temples were also engaged in installing and consecrating images in those temples (Raddock 2017). In the earlier periods, temples were mainly constructed in the forests, along the river banks, on hilltops or inside the forts, and on the seaside (Meister 1975–76, 1981, 2006). However, as time progressed, shrines were mainly designed for construction in town spaces, particularly at the end of the streets. The description aligns with the suggestion by the community of scholarship that temple styles and designs were prescribed for construction on the mountains as illustrated by temples like temples such as Chandrabhaga, Chalukya, and Ellora (Tarr 1970). The nature-proximate spaces were mainly preferred since they were considered holy, and they provided worshippers with a favorable space to connect with their gods (Neubauer 1981). The classification here was geographical context-based relating to the naturalistic settings where the temples and shrines were being constructed.

A seminal text on temple design, Devalaya-Vastu (Rao and Vikhanasacharyulu 1997), presents the tenets of temple design. The texts do not offer any distinctive practices on site selection, examination, and construction of shrine, based on style whether it was the North India centered, Nagara; Southern India focused, Dravida; or as prevalent in the East, Vesara. The description of various styles of temple architecture is well documented in writings of Acharya (1927a, b), Meister and Dhaky (1999), and Hardy (2012). Most discussion pertains to the "akara" or form of the "temple" though. Bharane presents a morphological expansion of temples, considering prakara of the temple, and sketches out the transformation of wayside shrines into temples in present day and age (Bharne and Krusche 2012).

So while there is typology (sometimes discussed as "style" in published discourse) on "akara," there is no discussion on typology of the temple precinct, "prakara", and "parisar" precinct in the temple design discourse building on classic canonical texts (Table 3.2).

The information reveals the typology of temples in terms of layout or plan organization of its built footprint, whether it is a singular shrine as a singular temple, example, deva-kulika or mandira, a group of collocated shrines as a temple, example, devalaya or dev-niketa mandala, or a network of temple spread through the town, example dev-niketa mandala or puri. The information also reveals the catchment context of the temple, whether it is a singular residence as in deve-kulika or mandira, a neighborhood as in devalaya or dev-niketa mandala, or a town as in dev-niketa mandala or puri.

Table 3.2 Information of temple as reference for typology

Sanskrit terms used for temple	Typological inference
Devayatana	Temple or a shrine—or does it include anywhere that god resides
Devalaya	Could be an altar within a home, fort, town—be referred to as an Devayatan (while temple always has a shrine, a shrine does not always have a temple built around it) No clue to size or scale
Mandira	References to scale: small to large, where small refers to shrine independently or shrine in a room, and medium refers to Temple or a building with a shrine: Temple or a shrine, a building or a room with a shrine
Deva-kula (-ika)	
Deva-niketa-mandala	References to scale: large to extra-large (a group of 2 to more)
Puri(-i)	References to scale: medium to extra-large, where medium refers to temple or a building with a shrine, large refers to larger footprint of the temple but in a marked compact location, and extra-large refers to footprint of the temple with an expanded, spread out footprint of the temple and shrines through a sizeable area of the town (the size of one to multiple neighborhoods?, check the scale of prakaras of current temples through study of figure ground plans) marked by gateways through the town
Mandara	This is more of a building typological reference and corresponds with Mandira architectural form

Design Syntax

Design syntax (Sharma 2015) is an approach to decodify the organization of a designed construct in terms of its composite design elements. This allows for re-configuration of the design elements to create new organizational plan patterns, volumetric forms, and designs if needed. The table below offers a synthesis of temples' design syntax based in Acharya's (1927a) Hindu dictionary description on Hindu temples. The information is decodified in terms of a design syntax of the temple, including prakara—the temple precinct (Table 3.3).

It should be noted that the temple style as in Nagara, Dravida, or Vesara was not deducible exclusively from these definitions and examples. The discussion on design syntax can be conducted irrespective of the specificity of architectural style.

Table 3.3 Decodification of design syntax of Hindu temple from textual taxonomic descriptions

Sanskrit terms used for temple	Synthesis of design syntax
Devayatana	Temple or shrine + water reservoir + gardens
Devalaya	Small temple: Garbhagriha/shrine + antarala/corridor + pradakshina path/perambulation path + ardhamandapa/front portico Large temple: Garbhagriha/shrine + antarala/corridor + 3 to 4 (pradakshina path/perambulation path) + 3 to 4 ardhamandapa/front portico
	Elements of a Deavalaya/temple: Precincts/prakara + gopura/gateway adorned with golden kalasas + tower/vimana + the great deity/maha linga + shrine/garbhagriha + main entrance/mahadvara + colonnade/kaisaleyalli Layout: garbhagriha/main shrine and surrounding shrines: (family) goddess shrine on left + (universal) goddess shrine on right + gopura at the main entrance/mahadvara + the colonnade/kaisaleyalli, to the south + the ancient images (puratana-vigraha) on the colonnade to the west + a row of lingas forming the 1000 (sahasra) lingas + colonnade + 25 pleasing statues (lilamurti, dhyana-murti) to the north + a separate temple (mandapa) to set up the god Narayana together with Lakshmi on the southwest side building
	Elements of broader temple context/includes reference to water: Temple + water tank/Kunda (could also refer to stepped well) + satra/institutional center + agrahar/village
	Elements of temple: Open mandapa on a base + a double row of pillars on the three exposed sides + roof of the ribbed dome standing on the 12 inner pillars + a large projecting porch on all three sides + principal mandapa + two inner rooms subset in the principal mandapa + shrine
	+ a prakara/enclosure + entrances + open mandapa Temple faces east and has three entrances on the north, south, and east, the east entrance, which is the main entrance, having two open mandapas at the sides inside

(continued)

Table 3.3 (continued)

Sanskrit terms used for temple	Synthesis of design syntax
Mandira	A type of rectangular building + encircled by a wall + broad and lofty tower + ranga-mandapa raised on a collection of beautiful stone pillars and adorned with rows of spouts + a car like the Mandara mountain + broad roads round the temple
Puri(-i)	A temple, an adytum, a building, a town

Chapter 4
Shiva Temples

This chapter provides locational and religious historical context to the 12 Shiva temples—Dwadash Jyotirlingas, further studied design syntax, landscape, urbanism, and ecology in subsequent chapters. The narrative draws heavily from popular media to get a sense of religious, cultural history associated with the built constructs of these temples, which is in a suspended state and, a consequence of a series of European, Moghul, and British, intrusions through the Indian subcontinent ranging approximately from 327 BCE to 1947; see Appendix IV for the Internet references that were reviewed to get a cross-sectional, anecdotal understanding on the topic.

Lord Shiva is considered one of the Tridev or Trinity in the Hindu religion. Brahma is the architect, Vishnu is the sustainer, and Shiva is the regenerator, the time itself, the first Yogi, and the detached one. The linga—an ovoid form—has been used through centuries to depict the eternal source, light, and endless time to represent the formless, nirguna, nirakaara Lord Shiva, also worshiped in sculptural forms, created by artists, based on best of their capabilities. The Jyotirlinga temples and shrines are deemed as the sites where Lord Shiva revealed or manifested his "being-ness" for the good of the devotees, denoted by a lingam. Each of the 12 Jyotirlinga sites has acquired denomination regarding the geographic site or associated legend of manifestation. The twelve Jyotirlinga are Kedarnath in Uttarakhand; Viswanath at Varanasi in Uttar Pradesh; Vaidyanath or Baijnath Jyotirlinga in Deoghar, Jharkhand; Somnath in Prabhas Patan, Gujarat; Nageswar at Dwarka in Gujarat; Mahakaleswar at Ujjain, Madhya Pradesh; Omkareshwar in Khandwa, Madhya Pradesh; Bhimashankar in Maharashtra; Grishneshwar at Aurangabad in Maharashtra; Triambakeshwar in Maharashtra; Mallikarjuna at Srisailam in Andhra Pradesh; and Rameshwar at Rameswaram in Tamil Nadu.

The narratives associated with religious significance and history of the temple inception for each of the temples are available as pamphlets, booklet, or postcards at each temple, under the title of temple history or Katha. Katha is a significant aural tradition of Vedic times where scriptures and knowledge were passed aurally and relied on being further disseminated similarly, thus relying much on memorization.

A. Sharma, *Mandala Urbanism, Landscape, and Ecology*,
https://doi.org/10.1007/978-3-030-87285-4_4

This allows room for human errors permeating or calling it editorial subjectiveness, not much different from translated and represented and re-edited works in print. Current prints of Shiva Purana and other books on the religious significance or association and period of the temple's inception in vernacular language were also reviewed, and scholarly articles were considered. Internet research was also used to check whether the popular narrative floating and influencing the World Wide Web and masses has any bearing on the printed narrative. Keywords of mythology and legend were plugged in the Google search engine to see how many results were generated; refer to Appendix IV for a list of results and relevant Internet sources from the first page. Despite the 100 thousand plus results, a closer review of the first page of the results showed that the content on most of those websites and blogs were reverberations of one or two critical narratives, also published in temple pamphlets. A narrative of anecdotal religious history is thus synthesized and presented below.

Kedarnath

Kedarnath temple is located at a height of 12000 feet in the Himalayan Range, in the Rudraprayag district in the Garhwal Himalayan range lying in the Uttarakhand state. The city is flanked by the Mandakini River on the west side, an 11-mile hike from Gaurikund. The last 22 kilometers (14 mi) of the temple access is not paved with roads for automobiles given the ecology and makes for an uphill trek from Gaurikund. Foot, ponies, and palanquins are the best means of transportation through this segment.

Due to extreme weather conditions, the temple is open to the public only between April (Akshaya Tritiya) and November (Kartik Purnima, the full autumn moon). During the winters, the vigrahas (deities) from the Kedarnath temple are carried down to Ukhimath and where the deity is worshiped for the next 6 months.

The Legend

The World Wide Web-based information network is galore with stories associated with the temple. Google search shows 316,000 results for legend and 413,000 for mythology, with many of these, as reviewed through results on the first three pages, being repetitious. The two critical legends related to the inception of the Shivalinga are shared below.

Legend 1: Nara and Narayana—two incarnations of Vishnu performed severe penance in Badrikashraya of Bharat Khand, in front of a Shivalinga fashioned out of the earth. Pleased with their devotion, Lord Shiva appeared before them and offered them a boon. Nar and Narayan requested Shiva to take up a permanent abode as a Jyotirlingam at Kedarnath so that all people who worship Shiva shall be freed from their miseries.

Legend 2: After the war of Mahabharata, five Pandava brothers walked toward Mount Kailash, considered as the abode of Lord Shiva, to seek the Lord's blessings to purge them of the sins of killing their cousins. Lord Shiva was unwilling to give darshans (viewing) to the Pandavas who fled Kashi to live incognito in Guptkashi. He turned into a bull and hid among the cattle on the hill. The Pandavas detected the Lord whereupon he tried to disappear by sinking himself head- first into the ground, but the Pandavas held him to seek his forgiveness. The forehead appeared at Pashupatinath in Nepal, the hump in Kedarnath, the two forelegs in Tungnath, the navel in Madhyamaheshwar, and the matted locks of Shiva appear in what is called Kalpnath or Kalpeshwar. The sites together are called Panch Kedar. Kedarnath is considered one of the holiest sites out of the five, perhaps because Goddess Sri Parvati is said to have meditated here for union with Lord Shiva, thus making it sacred in terms of both the masculine and the feminine energy of Shiva and Shakti. Another reason could be the correspondence of the location with the navel chakra in the yogic system, which is a critical link of prana or energy flow between the lower and upper chakras of the body.

Period of Temple Inception

Based on the legend, Pandavas have been said to build the temple or marked the site or shrine in Dwapara yuga, one of the yugas or period out of the four, in time classification per Vedic philosophy, and lasts 864,000 years. Another attribution is more recent and is associated with the Raja Bhoj of Malwa, who is said to have patronized the construction of the Kedarnath temple through his governance between 1076 and 1099 AD. Adi Shankaracharya (Indian philosopher and theologian) is mentioned as another significant patron of the temple in the eighth century.

Kashi Vishwanath

The Kashi Vishwanath temple is located on the banks of the Ganga River, in the city of Varanasi in the northern hemisphere of Uttar Pradesh.

The Legend

The Manikarnika Ghat on the Ganges banks near the Kashi Vishwanath temple is considered a Shakti Peetha, a revered place of worship for the Shaktism sect. The mythology of Daksha Yaga, a Shaivite literature, is considered an important literature that is the story about the origin of Shakti Peethas. It is said that Shiva came to the Kashi Vishwanath shrine through Manikarnika after the death of Sati Devi.

Period of Temple Inception

The legend stating Lord Shiva's passage by Manikarnika Ghat is indicative of the historic time period, which can be attributed to Satya Yuga. The legend of the Rameswaram temple mentions that Hanuman flew to get the longa from Kashi Vishwanath to bring it back to Lord Rama for offering prayers seeking forgiveness for killing Ravana upon his return from Lanka. This points to the existence of the lingam before the Ramayana time and incident, which is said to have occurred in the range of 3000–9000 BC.

Baijnath

Baijnath temple is in Deoghar town of District Jharkhand in Bihar state. It is one of the three temples that claim their shrines as "real" Jyotirlinga of Vaidyanath, among the other two in Parli, Maharashtra, and Baijnath, Himachal Pradesh. While the Dvadasalinga Smaranam has variation by which, the verse is Paralyam Vaidyanathan, i.e., Vaidyanathan is in Parli, Maharashtra. The names and the locations of the 12 Jyotirlingas mentioned are the following:

Saurashtre Somanathamcha Srisaile Mallikarjunam| Ujjayinya Mahakalam Omkaramamaleswaram || Paralyam Vaidyanathancha Dakinyam Bheema Shankaram | Setu Bandhethu Ramesam, Nagesam Darukavane|| Varanasyantu Vishwesam Tryambakam Gautameethate| Himalayetu Kedaaram, Ghrishnesamcha shivaalaye|| Etani jyotirlingani, Saayam Praatah Patennarah| Sapta Janma Kritam pApam, Smaranena Vinashyati||

Some sources identify Baidyanatham Chithabhoomau (1/21–24) and Sivmahapuran Satarudra Samhita (42/1–4) as ancient verses that identify location of Vaidyanath Jyotirlinga. According to these, Baidyantham is in "Chidabhoomi," which is the old name of Deoghar. The Government of India, Indian Railways, organizes a tour of Dwadash Jyotirlinga, which includes Baijnath Jyotirlinga from Deoghar. Thus, for the purpose of this study, Deoghar Jyotirlinga is being considered as one of the dwadasha Jyotirlinga.

Baidyanath Jyotirlinga temple, also known as Baba Baidyanath Dham and Baijnath, is a temple complex consisting of the main temple of Baba Baidyanath, where the Jyotirlinga is installed, and 21 other temples.

The Legend

According to Hindu beliefs, the demon king Ravana worshipped Shiva at the current site of the temple to get the boons that he later used to wreak havoc in the world. Ravana offered his ten heads one after another to Shiva as a sacrifice. Pleased with this, Shiva descended to cure Ravana, who was injured. As he acted as a doctor, he is referred to as Vaidya/Baid ("doctor"). From this aspect of Shiva, the temple derives its name.

Period of Temple Inception

The temple has been famous since the reign of the last Gupta emperor circa eighth century A.D. There is not much clarity about the year of the first construction of the temple. There is mention of the land and people of this tract and the shrine of Lord Vaidyanath at Deoghar in the Bhavishya Purana said to be compiled in the fifteenth or sixteenth century A.D.

Somnath

Commands a view from the tip of the Saurashtra peninsula, kissed by the waves of the Arabian coast.

The Legend

It is believed that Somraj, the moon god, built the Somnath temple or the Somnath Pattan out of gold. It was rebuilt by Ravana using the precious metal of silver as primary construction material.

Period of Temple Inception

Further, in the tenth century, it was rebuilt in stone by King Bhimdev Solanki. The other names of this famous Somnath temple are Deo Pattan, Prabhas Pattan, or Somnath Pattan. It was believed that the Somnath Temple was popular even in ancient times. Revenues were collected from 10,000 villages to maintain the temple. Given that Somnath temple was built of gold and silver, it was a constant target of looting and plundering by Muslim invaders from the North-West.

The present temple, built in 1951, is the seventh reconstruction on the original site. The temple has been constructed in the Chalukyan style with a shikhara nearly 50 m tall. The temple's imposing architecture includes intricate carvings, silver doors, an impressive Nandi idol, and the central Shivalinga. In the vast courtyard stand the massive mandapa (hall) and the main shrine, whose

gently curved pyramidal forms tower over the whole complex. One of the reconstructions is attributed to queen Ahilyabai Holkar in the eighteenth century.

Nageshwar

Nageshwar temple or Nagnath temple is located on the route between Gomati Dwarka and the Bait Dwarka Island on the coast of Saurashtra in Gujarat.

The Legend

This powerful Jyotirlinga symbolizes protection from all poisons. It is said that those who pray to the Nageshwar Linga become free of poison. The Rudra Samhita sloka refers to Nageshwar as "Daarukaavane Naagesham."

Daruka was an extremely cruel person of rakshasa pravrutti (evil, demonic nature) and often tortured the gentle population of Daruka vanam, but he was a devotee of Lord Shiva. Another Shiva devotee, Supriya, a merchant, reached the for trade, where Daruka lived with his wife, Daruki. Daruka asked Supriya to teach him further the ritualistic norms of performing pooja and penance to Shiva. Supriya suspected that Daruka might use enhanced powers gained through penance to cause harm and refused to take him as a student. This enraged Daruka, and he began to torture Supriya. Supriya, however, stood his ground and staking his faith in Lord Shiva, who eventually answered Supriya's prayers to manifest and kill the demon Daruka.

Period of Temple Inception

There is not much written about the construction of the Nageshwar temple through ancient times. There are some indirect references with regard to other temples. The Dwarkadheesh temple dedicated to Lord Krishna, located in neighboring Dwarka town, is a significant temple and is said to be surrounded by ancient temples from all four sides. One of these ancient temples could be the Nageshwar temple, which may have been worshipped as a shrine during the ancient times and may have been constructed and developed as a temple much later to preserve the shrine from direct contact with nature and visitors.

Mahaa-Kaleshwar or Mahakaal

The Mahaa-Kaleshwar temple is located in the city of Ujjain in the state of Madhya Pradesh, India. The temple is situated on the banks of the river Kshipra.

The Legend

Legend 1. According to an episode narrated in Puranas, a 5-year-old boy named Shrikar was inspired by Ujjain's King Chandrasena's devotion to Lord Shiva. Shrikar took a stone and started worshiping it regularly by considering it a linga. Others thought his act of worship was merely a game and tried to dissuade him but in vain. Pleased by the boy's devotion, Lord Shiva promised to reside in the forest in the assumed form of Jyotirlinga.

Legend 2. Once there lived a Brahmin in Avanti with his four sons; all were staunch devotees of Lord Shiva. Dushanan, a demon, used to disrupt their religious offerings and practice. The Brahmin and his sons invited brahmins from all over the land to pray Lord Shiva for a resolution; they created a linga form out of the mud to offer the prayers. The land where they had taken mud from, to make the linga, turned into a vast pond. When Dushanan came to disturb their pooja, Lord Shiva rose from this pond as Mahakaleshwar and destroyed Dushanan. On the request of the brahmins, Lord Shiva gave darshan to devotees at this site, which is also revered as one of the Shaktipeetha. Shaktipeethas are shrines that are believed to have been enshrined with the presence of Shakti empowered by the body parts of the corpse of Sati Devi, when Lord Shiva roamed carrying it, steeped in pain of her loss and inability to let her go, after her self-immolation due to her husband—Lord Shiva's insult by her father—King Daksha. Each of the 51 Shakti Peethas has shrines for Shakti and Kalabhairava. The Upper Lip of Sati Devi is said to have fallen here, and the Shakti is called Mahakaali.

Period of Temple Inception

The temple may have been constructed either before or during the times of King Ashoka, the last Mauryan emperor. This can be inferred because the city of Ujjain during Ashokan times was noted as a center of Jainism, Hinduism, and Buddhism, not expressly Shaivism. Ujjain was destroyed by Mughal King Iltutmish in 1235 and remained in Muslim control until 1750 when the Sindhus took control. The temple can be assessed to have undergone deconstruction and reconstruction through centuries of 1235 and 1750.

Bhimashankar

Bhimashankar is situated in the Sahyadri hills of Maharashtra. Bhimashankar temple is one of the well-known Jyotirlinga, among the 12 Jyotirlingas situated all over India. Bhimashankar is located in the village of Bhojgiri, 50 km northwest of Khed, near Pune. Bhima River (also known as Chandrabhaga River) flows in proximity and merges with the Krishna River southeast.

The Legend

Legend 1. Bhima was the son of Kumbhakarna, in the Ramayan times. When Bhima learned about his father's death in war with Lord Rama, an incarnation of Lord Vishnu, from his mother Karkati, Bhīmā was infuriated and vowed to take revenge against Lord Vishnu. He pleased Lord Brahma through penance and secured powers in blessings from him. Bhima abused the power, tortured the gentle population, and disrupted religious activities. He captured a staunch devotee of Lord Shiva—Kāmaroopeshwar—and put him in the and insisted that Kāmaroopeshwar worship him instead of Lord Shiva. Upon refusal by Kāmaroopeshwar, Bhīmā raised his sword to strike the Shiva Linga hand-built by Kāmaroopeshwar for offering prayers. However, as soon as Bhīmā managed to raise his sword, Lord Shiva appeared and reduced the evil demon to ashes, thus concluding the tyranny. All the Gods and the holy sages present at the location requested Lord Shiva to make this place his abode. Lord Shiva thus manifested himself in the form of the Bhīmāshankar Jyotirlinga.

Period of Temple Inception

The Bhimashankar temple has the sanctum at a lower level and is a composite of old and new structures and is dated around the thirteenth century. Additions are said to have been made in the eighteenth century by Nana Phadnavis. The great Maratha ruler Chhatrapati Shivaji Maharaj is said to have made endowments to this temple to facilitate religious services.

Omkareshwar

Omkareshwar is situated in the Khandwa district of Madhya Pradesh State, about 12 km from a small town called Mortakka. It is on an island called Mandhata or Shivapuri in the Narmada River. There are two main temples of Lord Shiva here, one to Omkareshwar, the Lord of Omkara or the Om sound, located in the island, and one to Amareshwar or Amaleshwara and mentioned as Mammaleshwar as well—the Lord of immortality, situated on the south bank of Narmada River on the mainland.

The Legend

As per the Shiva Purana, two sons of the sun dynasty Mandhata—Ambarisha and Muchukunda—practiced severe penance and austerities here and pleased Lord Shiva. They also performed great religious sacrifices in this place; subsequently, the mountain got to be known as Mandhata.

Another popular legend says that once upon a time, Vindhya Parvat practiced severe penance and worshipped Parthivarchana along with Lord Omkareshwar for nearly 6 months. As a result, Lord Shiva was pleased and blessed him with the desired boon. At the sincere request of all the gods and the sages, Lord Shiva made two parts of the lings. One half being Omkareshwara and the other Amaleshwara or Amareshwar.

Period of Temple Inception

The legend associated with King Mandhata gives a clue to the temple inception. The King belonged to the Ikshvaku clan, an ancestor of Lord Ram. These events give an indication of the temple's inception. Stories of King Mandhata were found to be carved in stone during the Mohenjo-Daro period, attributed to about 5000–2500 BCE, prompting some to estimate the period of King Mandhata and his sons to around 6200 BCE; thus, the shrine to Lord Shiva could be attributed to that time.

Triambakeshwar

Triambakeshwar (also spelled as Trayambakeshwar) temple is located in the town of Trimbak, in the Nasik District of Maharashtra state, 28 km from the city of Nasik. Kusavarta Kunda, or pond in the temple premises, is said to be the source of the Godavari River.

The Legend

Legend 1. The legend goes that a sage named Gautam Muni resided on the Brahmagiri hill with his wife, Ahilya. By virtue of his devotion, the sage received a blessing from Varuna, the God of rains, for rain in his ashram every day, thus ensuring a consistent supply of grains and food for all people and cattle in the ashram. The other rishis, jealous of his fortune, arranged for a cow to enter his granary and paddy fields. Gautam Rishi attempted to ward off the cow with a bunch of Darbha grass and inadvertently killed it, thus acquiring the sin of the cow's death. Gautam Rishi worshipped Lord Shiva to bring the Ganga down to his hermitage to purify the premises and absolve him of the sin. Pleased with devotion, Shiva requested Ganga to flow down and cleanse the premises and the Rishi and also to stay there eternally for the public good. At the request of sage Gautama and others, Lord Shiva agreed to reside by the river Gautami and hence the shrine at Triambakeshwar.

Legend 2. Another popular legend pertains to the Lingodbhava narrative around Lord Shiva. It says once Brahma and Vishnu searched in vain to discover the origin of Shiva, who manifested himself as a cosmic column of fire. Brahma lied that he had seen the top of the fire column and was cursed that he would not be worshiped on earth. In turn, Brahma cursed Shiva that he would be pushed underground. Accordingly, Shiva came down under the Brahmagiri hill in the form of Triambakeshwar. Triambakeshwar temple is the only place where Shivalinga is subsided in the floor.

Period of Temple Inception

Reference to Rishi Gautama can be used to get an idea of temple inception. The Rishi Gautama is mentioned in Ramayana, which is approximately dated circa 7323 BCE, so the lingam's appearance can be associated with that time. The temple was later said to have been reconstructed by Nanasaheb Peshwa Balaji Baji Rao.

Ghrishneshwar

This pilgrimage site is located in Ellora (also called Verul), less than a kilometer from Ellora Caves—a UNESCO World Heritage site. It is about 30 kilometers northwest of the city of Aurangabad and roughly 300 kilometers east-northeast from Mumbai.

The Legend

According to Shiva Purana, in the southern direction, on a mountain named Devagiri lived a Brahmin called Brahmavetta Sudharm along with his wife, Sudeha. The couple did not have a child, because of which Sudeha was sad. Sudeha prayed and tried all possible remedies but in vain. Frustrated with being childless, Sudeha got her sister Ghushma married to her husband. On her sister's advice, Ghushma used to make 101 lingas, worship them, and discharge them in the nearby lake. With the blessings of Lord Shiva, Ghushma gave birth to a baby boy. Because of this, Ghushma became proud, and Sudeha started feeling jealous toward her sister. Kusuma used to pray Shiva daily, ritualistically, immersing a Shivalinga in a tank. Her husband's first wife, envious of her piety and standing in society, murdered Kusuma's son in cold blood and threw him in the lake where Ghushma used to discharge the lingas. An aggrieved Kusuma continued her ritual worship, and one day when she immersed the Shivalinga again in the lake, her son was miraculously restored to life. Shiva is said to have appeared in front of her and the villagers; Ghushma asked Lord Shiva to forgive Sudeha and emancipate her. Pleased with her generosity, Lord Shiva asked her another boon. Ghushma asked him to reside at the location eternally for other devotees' benefit and for the linga to be known by her name.

Period of Temple Inception

The Mughals destroyed this temple during invasion of the thirteenth and fourteenth century. The temple went through several reconstructions during the Mughal-Maratha conflict. The temple was reconstructed by Maloji Bhosale of Verul (grandfather of Shivaji) in the sixteenth century and later by Indore's queen Ahilyabai Holkar in the eighteenth century.

Mallikarjuna Srisailam

The temple is in the mountain range of the Eastern Ghats, known by different names like Srigiri, Srimala, Sringeri, and Rishabagiri. Srisailam is a town in the state of Andhra Pradesh.

The Legend

Rishaba or Nandi Deva did penance on this hill and obtained the darshan of Lord Shiva and Devi Parvati, hence the name Rishabagiri. Shiva is worshiped as Mallikarjuna and his consort Parvati as Brahmaramba. The Jyotirlinga is thus also a shaktipeeth.

Legend 1. Long ago, Princess Chandravathi of the Chandragupta dynasty faced a domestic calamity and decided to forsake royal comforts. She went to the Srisailam forests and lived on fruits and cow's milk. One day, she noticed that one of the cows was not yielding milk. Later she learned through her herdsman that the cow was going to a secluded spot and showering milk on a linga amid mallige (jasmine) creepers. The next day she went to that spot and witnessed the miracle. The same night Lord Shiva appeared in her dream and asked her to build a temple at this spot. Since the linga was entangled in mallige creepers, the deity was named Mallikarjuna.

Legend 2. According to another legend, Lord Shiva once came to the Srisailam forest for hunting. There He met a beautiful girl of the Chenchu tribe, fell in love with her, and decided to stay in the woods. The girl was none other than Goddess Parvati Herself. In the temple, there is a bass relief depicting this story. Even today, people of the local Chenchu tribe are allowed into the sanctum, a privilege typically reserved for the temple Priest. On the night of Maha Shivaratri, they are permitted to go into the sanctum and perform abhisheka and puja.

Period of Temple Inception

Princess of Chandragupta dynasty gives a clue to the initial discovery of the lingam Chandragupta Maurya dynasty is said to have been ruling through 321–297 BCE, thus bearing reference to the discovery of the lingam and preliminary construction around the shrine.

Rameswaram

Rameswaram (also spelled as Ramesvaram, Rameshwaram) is a town in the Ramanathapuram district in Tamil Nadu state. The temple is on Pamban Island (also known as Rameswaram Island) in the Gulf of Mannar, at the tip of the Indian peninsula connected to mainland India by the Pamban Bridge.

The Legend

The Jyotirlinga was worshipped by Lord Rama to atone the sin of killing Ravana. Hanuman flew to bring the linga from Kailasa (or Benaras/Varanasi), for Lord Rama to worship. As it was getting late, Rama worshipped the lingam made of sand by Sita Devi. This lingam worshipped by Lord Rama is known as Ramanathar. When Hanuman returned, he was disappointed that his Lord had not used the lingam that he had brought. Lord Rama pacified Hanuman and named this lingam Kasi Viswanathar. Devotees should worship Kasi Viswanathar before worshipping Ramanathar.

Period of Temple Inception

The temple's inception can be dated to Ramayana time in view of Rama's return from Sri Lanka. Since Ramayana events are estimated to have taken place between 3000 and 9000 BCE, and his return from after the conquest of Ravana, to around 7335 BCE, this can be predicated as the founding period of the Rameshwar temple.

Chapter 5
Temple Town Urbanism, Landscape, and Ecology

Design Syntax: Mandir and Prakara

The mandir or temple building footprint shape conforms with the auspicious shapes per Vastushastra rules in all cases. See figure ground study illustrating the same in Appendix III. The Prakara—the premise of the temple—however, does not reflect a similar consistency of conforming with the auspicious shapes. Some prakaras, such as in Omkareshwar temple, inadvertently yield undesirable (inauspicious) shapes, due to incremental iterative construction without clear plan and considerations of Vaastu purusha mandala. Given the intense construction and building activities over centuries, it is difficult to assess the original soil quality and natural land slopes, contiguous to the temple, and thus expectant changes to those in current times. It is safe to speculate that the temples were most likely built on Gaja prustha/elephant back or Koorma prustha/turtleback, since access to them requires climbing up in the east-west direction, and these are forecasted as a good landing for enthusiasm, wealth, and well-being.

The design syntax of the current day urban or peri-urban temple can be synthesized in terms of Mandir + Prakara + Parisar: Temple + Plaza (tree + benches + water fountains) + Kiosks in the context beyond the temple perimeter.

Layout Mandala of Temple Towns

The geographical context of the temples has been stated as mountain, sea, or river. The contextual orientation to the temple site is helpful before zooming into the town layout and is presented in Table 5.1.

The study of mandala layout of the temple parisar or the area right outside the temple prakara (precinct) reveals the variety of spatial patterns. Dandaka layout has

A. Sharma, *Mandala Urbanism, Landscape, and Ecology*,
https://doi.org/10.1007/978-3-030-87285-4_5

Table 5.1 Geographical context of the temple

Shiva temples	Physical/built environmental context
Kedarnath, Uttarakhand	Extreme mountainous glacier valley
Vishvanath, Varanasi, Uttar Pradesh,	Riverside
Vaidyanath, Deoghar, Bihar	Inland, in proximity of river
Somnath, Saurashtra, Gujarat	Seaside
Nageshwar, Dwarka, Gujarat	Seaside/inland, in proximity of a gulf
Bhimashankar, Maharashtra	Mountainous plateau
Ghrishneshwar, Maharashtra	Inland, in proximity of river
Triambakeshwar, Maharashtra	Inland, in proximity of river
Mahakaal, Ujjain, Madhya Pradesh	Inland, in proximity of river
Omkareshwar, Khandwa, Madhya Pradesh	Riverside
Mallikarjunam, Shri-Shailam, Andhra Pradesh	Mountainous plateau, in proximity of river
Rameshwaram, Tamil Nadu	Seaside

a simpler grid layout and is said to be associated with Vanaprastha—a transition from the worldly to the spiritual stage of a life. It has a quadrangular shape and is prescribed to have 3–5 carriage (secondary/local highway) roads, and minor (tertiary/collector) roads from east to west are optional, so is to have a prominent central axial (primary/state highway) roadway; a ditch surrounding the settlement further enclosed with a wall with gates. A temple of Vishnu should be built on the westside, in Varuna or Mitra part of the Mandala, and a Shiva temple on outside the town or within village perimeter in the Parajahya (northeast) and Udita (northwest) parts, but always on the north direction (See Acharya 1927a, b, pg 10, for Mandala with sub-grid references). Kedarnath temple projects a Dandaka layout, not in the pure sense of conformation with precise details of a number of roads, but approximately, formally. Chaturmukha and Nandyavarta mandalas have the temple located in the center of the grid, with an inside-out expanding quadrangular form and are thus ruled out. Prastara and Padmaka both have more complex subdivisions within the two predominant quadrangles and an inside-out expanding grid. Svastika and Sarvatobhadra illustrate an intermediate variation of grid ranging between the simplicity of Dandaka to the complexity of Prastara and Padmaka. None other thus comes close to the diagrammatic projection of the parisar of Kedarnath and Omkareshwar and then Dandaka (Figs. 5.1 and 5.2).

The Kashi Visvanath temple has a Sarvatobhadra mandala, which can be either a rectangle or a square planning template with inside quadrangles planned for residences and the ones next to carriageways for trade-related transportation. Sarvatobhadra mandala can be circular or quadrangular and has axial carriageways running through the layout connecting from north to south and west to east (Fig. 5.3).

The Triambakeshwar temple has an off-centered Padmaka, Knarvata, or Nandyavarta Vartul surrounded by possibly a Prastara, Sarvatobhadra, or a polyaxial multiples of Dandaka, with the periphery projecting a composite of the Padmaka and the mandala that surrounds it. The center, however, seems to be closer in resemblance with Knarvata, given a prominent sense of center and the circular (or in this

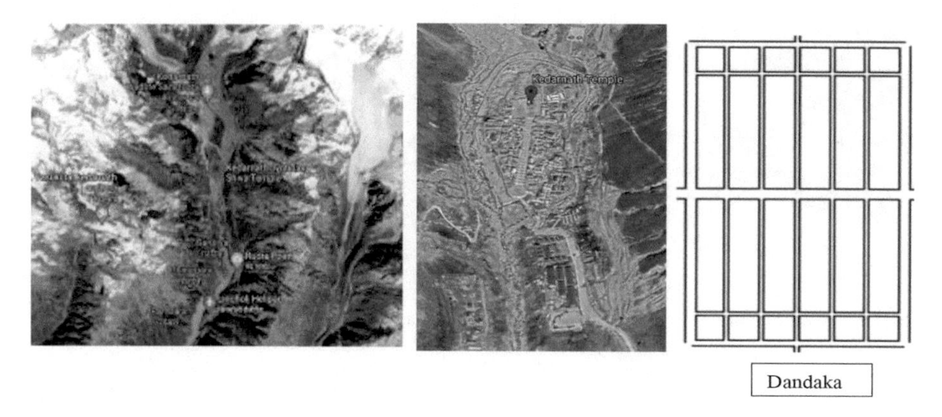

Fig. 5.1 Geographical context and Mandala- Kedarnath, Uttarakhand

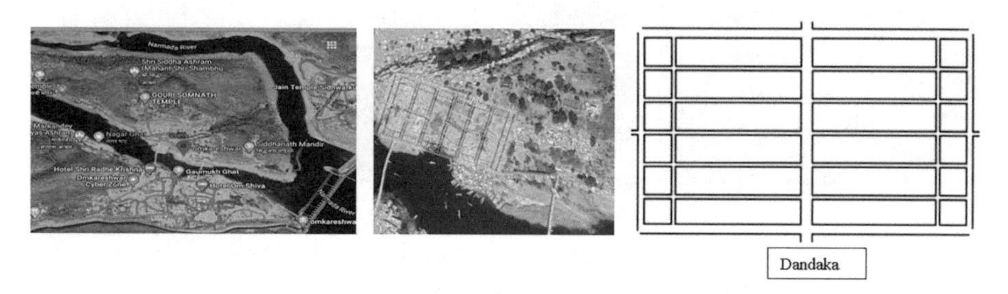

Fig. 5.2 Context and Mandala- Omkareshwar, Madhya Pradesh

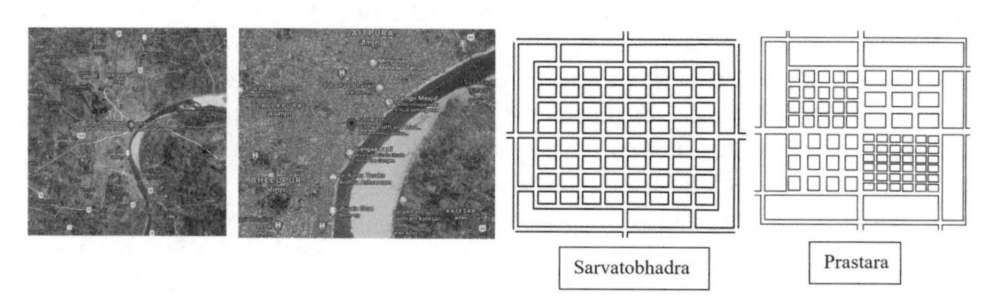

Fig. 5.3 Context and Mandala- Kashi Vishwanath, Uttar Pradesh

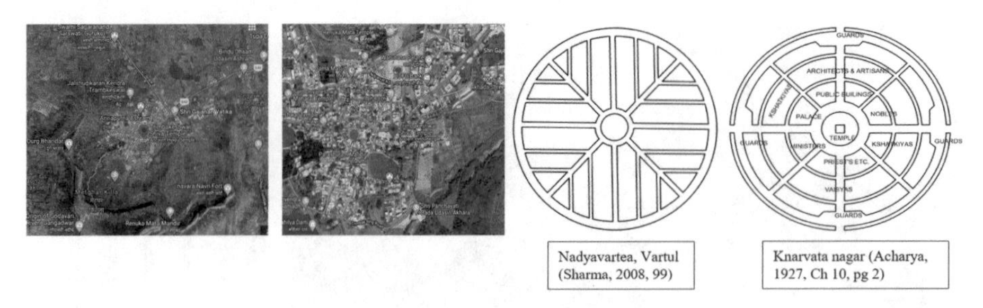

Fig. 5.4 Context and Mandala- Triambakeshwar, Maharashtra

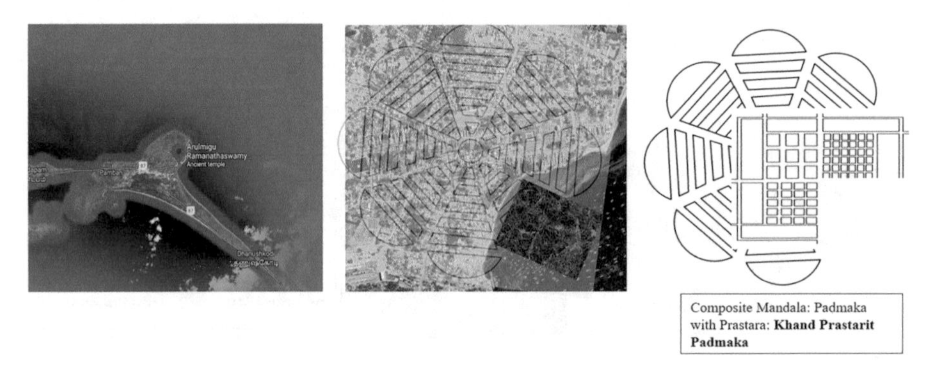

Fig. 5.5 Context and Mandala- Rameshwaram, Tamil Nadu

case, quadrangular) grid radiating outward with a demarcation on crucial access points or carriageways both north-south and west-east (Fig. 5.4).

The Rameshwaram parisar is challenging to categorize as it is located at the seaside and yet does not have the radial semi-circular layout as Karmuka, perhaps closer to Kheta or Khetaka, but again does neither have radial settlement pattern nor a walled perimeter and nor does it have a mid-land, half polygonal grid characteristic of a Kubjaka. Looking from a distance, the perimeter of the town seems more like Padmaka, with the edge form containing at the same time expanding outward. It can very well be read as the composite Mandala, a Padmaka with Prastara at the center or Prastarit Padmaka, just as in Baijnath (Fig. 5.5).

Baijnath town projects a complex Mandala with the Padmaka form reflected at the periphery, with the edge form containing at the same time expanding outward but containing a Prastara layout within the circular boundary (Figs. 5.6 and 5.7).

Nageshwar presents a unique context with just the built footprint of the temple surrounded by farmland, with small groups of houses dotting the landscape sporadically to project any strong pattern, but an amorphous fabric. From a bird's eye view, it is possible to see a sizeable housing density, further north of the temple. The housing layout is loose in form with another local temple at the center. Given the id-land

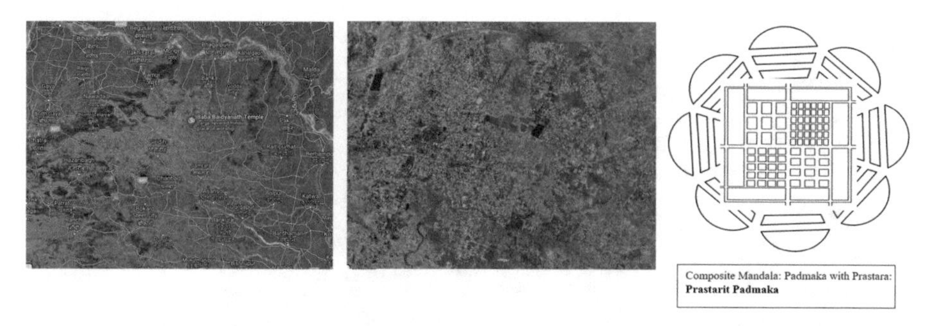

Fig. 5.6 Context and Mandala- Baijnath, Bihar

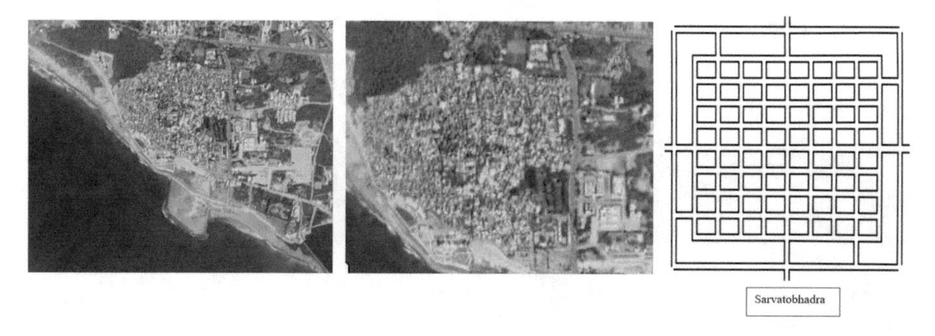

Fig. 5.7 Context and Mandala- Somnath, Gujarat

context of the temple and temple at the center, and thus the projection of the ideal that the village could have a Knarvata form, it is being diagrammed as such in Fig. XX. Further zooming out, it is possible to see another relatively lighter housing density to the southwest of the temple, laid out in what can be best assessed as the preliminary beginning stage of the Dandaka form. The resultant urban mosaic sets out a network of (currently, loosely applied) mandala forms (Fig. 5.8).

Bhimashankar, located on a rocky plateau, has a small settlement in the proximity, an embedded part of the economic and religious ecosystem, around the temple. The residential settlement has a simple Dandaka form, which follows the ridge topography. Ghrishneshwar lies on flatter topography by the riverside, with the town pattern closely resembling Padmaka or Sarvatobhadra (Figs. 5.9 and 5.10).

The neighborhood around the Mahakaal temple project a complex mandala of multiples of Sarvatobhadra with a Dandaka running through it (Fig. 5.11).

Mallikarjun temple town resembles a cross between Svastika and Nandyavarta chaturastra. I can also be viewed as multiple Dandaka on the right-hand side of the temple and half a Prastara on the left-hand side of the temple (Fig. 5.12).

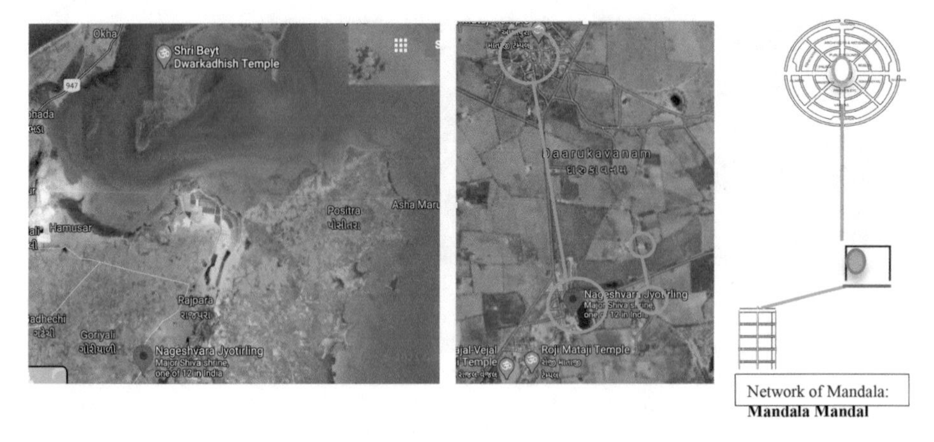

Fig. 5.8 Context and Mandala- Nageshwar, Gujarat

Fig. 5.9 Context and Mandala- Bhimashankar, Maharashtra

Fig. 5.10 Context and Mandala- Ghrishneshwar, Maharashtra

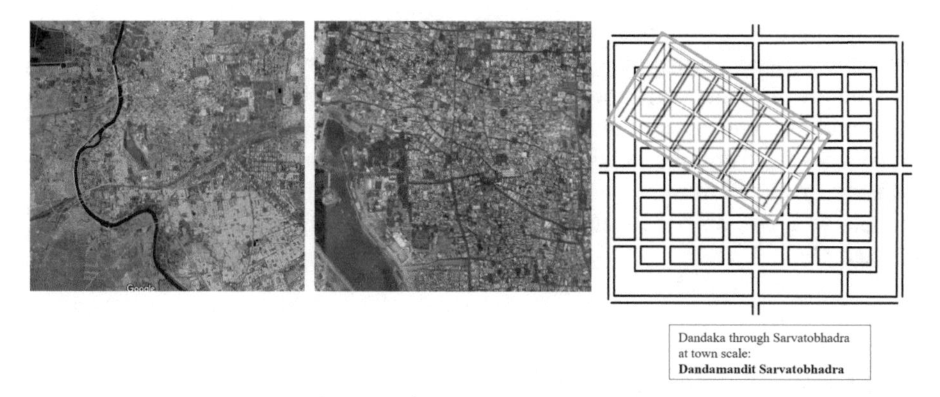

Fig. 5.11 Context and Mandala- Mahakaal, Ujjain, Madhya Pradesh

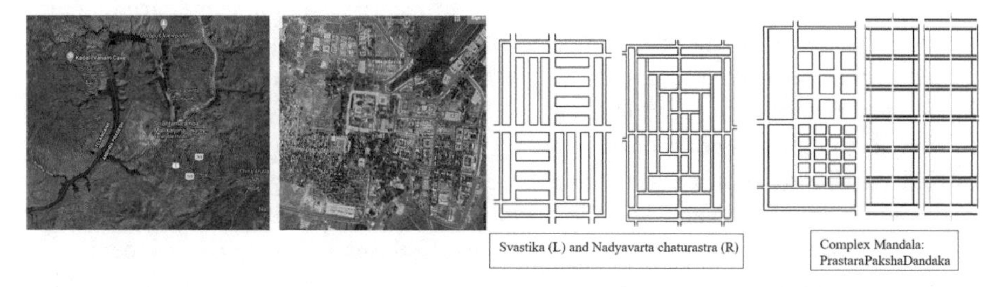

Fig. 5.12 Context and Mandala- Mallikarjunam, Andhra Pradesh

Access and Processional Pathways

The streets within the mandala and access pathways to the temple facilitate and enhance the processional experience of temple visits. The temple visits are a part of practicing or evoking Bhakti/devotion to God. Bhakti is a way to praise. It also is a preliminary step in the journey to understanding God. The "procession"/yatra and pradakshina are all a part of this process from walking from one's being/place to God/symbolic sthal of residence. The temple of pilgrimage for Hindus is a symbolic place where God resides, and thus these places need to be taken care of daily, through rituals by the local priests and prayers by temple visitors. Hindu belief of God otherwise is Nirankara (formless) and Sarvavyapi (omnipresent).

In contrast, the expressions as Shiva and Shakti were perhaps to communicate messages of Bhakti and Nirvana in physical form to a community, predominantly accustomed to the material realm. The procession through the access pathways has an element of sound in chants called jaykaras, which is not captured in these images. The procession thus is tightly woven in clusters of people, the group varying from small to large, when people are wanting to experience that sound energy outwardly while processing the spiritual communion inwardly. Sometimes people trail off, and you see families or couples or individuals in there when they take a break from the social experience of the procession to reflect in their own unit or quietness.

The processional paths through rugged terrain or topography in the parisar, the surrounding context, are guided by topography and have duly located protection barricades to ensure the safety of visitors. The barriers act as a control mechanism to keep the crowd movement directed and controlled within the prakara and to keep people at a safe distance from the sacred shrine located in the maha-mandapa of the temple.

Congregation

The social preference of spaces within prakara and parisar is primarily guided by the purpose of the visit to the temple. Most visitors arrive for darshan/spiritual communion through viewing or pooja/performance of rituals. These two functions are either being undertaken as a part of teertha yatra—pilgrimage or as just yatra or travel, which is religion-centered (Fig. 5.13).

The social congregation near the water at the Mandakini River in the proximity of Kedarnath is very different to the congregation by the Ganges river ghats close to Vishwanath temple in Varanasi. The primary difference here is the purpose and significance attached to those rivers and temples. Kedarnath Shiva linga's inception is associated with Shiva's passage through the terrain. So is Kashi Vishwanath, which is associated with Shiva's passage in an aggrieved state and thus energetically charges state of his wife Sati's loss with her dead body, thus lending the place an enhanced and combined energy of both masculine and feminine, of living and the dead, of perseverance of the living through releasing of the dead. The sending off the deceased on their onward journey is marked with rituals, which maybe more for a personal emotional release in coming to terms with the departure. The ritualistic release is conducted in witness of the eternal natural elements of fire, returning the composite elements—the panch mahabhoot (the five elements of agni, jal, vayu,

Fig. 5.13 Congregation configurations at temples. Kedarnath (L), Baijnath(R)

akash, and asthi: fire, water, air, space, matter) of the human body back to nature. The ash dispersal in the flowing river symbolizes the passing over to the other realm (to continue onward journey through rebirth or an absolution from such through nirvana). The Mandakini River which flows by Kedarnath temple flows into Alakananda, which then passes by Vishwanath temple, where it is known as Ganga, and acquires a venerable reverence as connected to nirvana through association with Vishwantah temple's religious narrative.

The congregations next to water are for leisure and contemplating nature or spirituality. These are common in Omkareshwar Narmada river and Rameshwaram. Many temples are located in the proximity of the water, where the ritualistic ablution or dip in the Kunda water occurs within the Prakara or temple premise and nature-gazing in the Parisar/proximate context at the river. While ritualistic dip or ablution in the holy water to cleanse yourself is mandatory and strictly followed before the darshan in some temples such as Rameshwaram, it is optional for others such as Omkareshwar which has no rigid prescription regarding ablution before or after the darshan.

The social congregations are also guided by a few other existential realities such as restrooms, resting space or benches, and shopping, be it for prasad/offering to Gods or souvenirs.

Temple parisar/town context typically hosts the built infrastructure for an economic ecosystem that is generated through temple tourism. The economic activities range from more formal such as the permanent structures such as hotels, resorts, lodges, and samaj/community-organized atithi bhavans/guest houses to more temporary ones such as camping tents, shops, and retail stores to informal moveable flower, coconut, and pooja material vending kiosks. Over the years, a network of smaller, newer shrines pops up along the route or pradakshina of the main historical temple. Tourist or devotees' attitude to this network of shrines is reflected through the congregation patterns as well, which is not of higher intensity or duration, indicating that sometimes these smaller satellite networks of shrines and temples, especially in places of historical significant site/temples, can be viewed as annoyance and unwanted aberrations.

The narrative of the social congregation is integrally weaved in the ecology of temple tourism as well. According to the Ministry of Tourism, 22 million leisure tourists visited the Madhya Pradesh region housing Omkareshwar temple between 2011 and 2012. Approximately half this number—8.5 million non-leisure tourists—were recorded in this state around the same time. The most popular time visited by non-leisure tourists was during October, while the number of non-leisure visitors spiked in April. The statistics on tourism are for Madhya Pradesh attracting the maximum number of tourists from Delhi, Uttar Pradesh, Chhattisgarh, and Maharashtra.

International tourism records show roughly 21% visitors from the United Kingdom and 14.7% from the United States. Madhya Pradesh has several architecturally and religiously significant temples including Kandariya Mahadev Temple, Matangeshwar Temple in Khajuraho, and Chaturbhuj Temple in Orchha palaces, besides sanctuaries, stupas, forts, archeological sites, indigenous cultural fairs, and festivals[14]. Jharkhand received 104.19 lakh visits by domestic tourists, and same-day visitors were made in Jharkhand at the 59 popular tourist destinations from July 2009 to June 2010. In addition, foreign tourists and same-day visitors made a total of 15,557 visits to these destinations. Tourist traffic followed a seasonal trend in Jharkhand. Tourism peaks in January–February, with foreign tourists peaking in August or December. However, almost half of the domestic tourists stay with friends and relatives and others at hotels. At the same time, 90% of the foreign tourists stay at hotels. Hotel guests peaked in January–February. Most foreign tourists come from the United States, United Kingdom, Europe, and Germany, and most domestic visitors are from Jharkhand, West Bengal, and Bihar. Tourists from these states made up 90% of the domestic tourists (3). Varanasi, where Vishwanath temple is located, received 65,786,949 domestic tourism and 1,907,399 foreign in 2017. Tourism destinations include Sarnath, a Buddhist religious site; Ganga Ghat, Ganges River waterfront; and Kashi Vishvanath Temple, Hindu Temple. Despite its tough terrain and temple and access closure most of the year, Kedarnath receives about 694,934 pilgrimage tourism and was estimated to reach one million with rise in air taxi service.

The urbanism and ecology created through the collusion of the social, the physical, the economic, and the spiritual is most pronounced in temple towns, generating unique temple urbanism and ecology. The reference also connects to the four phases and purposes of human life as prescribed by Vedic texts. These are Dharma, Artha, Kama, and Moksha, implying the duty-centered, economy- or revenue generation-centered, physical pleasure-centered, and liberation-centered. The temple parisar and towns are an excellent example of where these four purposes co-occur, interwoven with the next, parallel to each other. Almost a reflection of contemporary society, which does not practice the Vedic prescribed phasing of sequential focus and living in that order but all or select few of the tenets practiced as compressed in a day or a week or month or another individually preferred capsule of the time and repeated over and over. The economic ecosystem centered around the temple or the temple visitor has supplemental loops of interconnections and support being generated. The temple and visitor relationship inadvertently generates a circular and semi-open economy loops through a donation to the temple and through the transaction with vendors. Despite the interconnected existential loops, the essential spirit of the Hindu temple has been retained in terms of a place for conducting rituals, a quiet pause for contemplation or meditation, an outlet for the performance of daan-dharma, or donation-duty of supporting others through transactions with local vendors, prasad kiosks, or donation boxes.

Landscape and Ecology

Most temple towns are in highly urbanized locations, barring those in harsh climatic terrains such as Kedarnath, Nageshwar, Triambakeshwar, and Mallikarjun. Visual reconnaissance of maps from bird's-eye view indicates the built area-to-green cover ratio, as of 80:20 percent, as an average. Vastushastra offers a guide for temple materials. The material palette of the temple town sets the tone for the look but also in terms of ambient temperature. In an intensely built environment, the selection of materials can make a difference of up to 3 degrees to the heat island effect. Materials are used in different combinations, but they largely comprise local stone (sandstone/limestone/basaltic), timber, and concrete and brick masonry for iterative construction and development.

A lighter footprint of lighting was commonly observed in all temple prakaras. This is to preserve any damage to temple material and structure. The soft lighting was not practiced in the town, because pooja and food vendors always used it as a tool to outbid each other with loud lighting to attract customers. Yet, there is a need for considering the lighting as a design element to supplement the experience of the processional space to the temple, walking outwards-in, during late night/shayan aarti, and early morning, mangala aarti and darshans. The temporary lighting arrangements for the commerce in the temple Parisar can be damaging to the temple and detrimental to the visitor's experience.

Tree canopy is very patchy in many temple prakaras such as Triambakeshwar and Nageshwar but also scant and scattered in parisars such as for Rameshwaram, see Fig. 5.5. The four Mahadwaras marking the access to the temple prakara/premise have not been rendered any additional emphasis through landscaping. The roads bounding the temple premise and main thoroughfare roads are also devoid of green canopy. Temples in the urbanized context.

Designed landscapes are introduced within the temple prakara, mainly to provide shade in resting places or to reduce the heat and provide a cooling effect for the premise.

However, these landscapes end up looking like cosmetic add-ons, as the scale, spatial organization, formal patterns, and selection of plant materials are not reflective of the ancient period that the temple architecture represents. The mismatch is jarring, especially when the region is rife with the local biodiversity and rich vegetation mix.

Water Features

Most of the studied temples are located near a river such as Kedarnath close to Mandakini, Triambakeshwar to the Godavari, and some right on the riverbank such as Omkareshwar on Narmada and Vishwanath on Ganga. Rameshwaram and Somnath are located right on the coast. The 12 temples have a Kunda, a water feature, within the temple premise, as named in Table 5.2.

Most temples, if they are not located by a water body such as an ocean or a river, have either a natural or constructed water body in the vicinity, such as Nageshwar and Triambakeshwar. The water ponds in the vicinity of temples, which are not integrated as a part of the ritualistic cycle, are suffering from pollution and neglect, filled with an algal bloom, such as the Rudrasagar near Mahakaal.

Table 5.2 Shiva temple towns and water features

Shiva temples	Temple related Kunda/water tank in Prakara	Temple related Kunda /water tank in Parisar	River/ocean in Parisar/ proximity of 1 mile	Closest river or stream at a distance, beyond 1 mile
Kedarnath, Uttarakhand		Gaurikund (hot water spring)	Mandakini river	
Vishvanath, Varanasi, UP			Ganga river	
Vaidyanath, Deoghar, BIHAR		Shivaganga		Ajay
Somnath, Saurashtra, GUJ			Arabian Sea	Hiran and Kapila rivers
Nageshwar, Dwarka, GUJ			Gulf of Kutch, Arabian Sea	
Bhimashankar, Maharashtra		Mokshakund	Bhima River	
Ghrishneshwar, Maharashtra	Shivalaya Tirtha Kunda		Velganga River	
Triambakeshwar, Maharashtra		Kushavarta Kunda	Godavari River	
Mahakaal, Ujjain, MP		Rudra Sagar (not related to temple)		Kshipra River
Omkareshwar, Khandwa, MP			Narmada river	
Mallikarjunam, Shri-Shailam, AP				Krishna River
Rameshwaram, TN	22 Kunda (freshwater springs) on Oceanside		Palk Strait, Indian Ocean	

The current landscape design within the temple prakara, parisar, and town does not project the vegetational character of the region, as summarized below. Kedarnath is located at a high altitude of 3583 meters above sea level, making the winter extremely cold with a low of at least 32' Fahrenheit and an annual rainfall of 1475 mm. Both the terrain and climate are thus quite dramatic and extreme. The vegetation is mainly coniferous, alpine, subalpine, meadows, growing rhododendrons, oak, birch, and pines. Faunal life is rich with avian, mammal, and reptile diversity. Some mammals found in this region are Indian jackals, the Himalayan black bear, and the snow leopard. Reptiles such as Himalayan pit viper and the water snake Boulenger's keelback and birds like rusty-flanked treecreeper, little pied flycatcher, and the Uttarakhand state bird—Himalayan monal—also inhabit the region. Baijnath is a semiarid landscape, with a climate that is mild, and generally warm and temperate with average annual temperature of 25.4 °C (77.72 F) with the highs of 31.8 °C (89.24 F) in May and lows of 17.3 °C (63.14 F) in January, annual rainfall of 1174 mm (46.22 in) with most rains in monsoon through July, averaging 298 mm (11.73 in). The Jharkhand flora thus mostly consists of dry and moist deciduous forests. About 200 species of avifauna are found in the territory of Jharkhand. The forested areas, including the national parks and wildlife sanctuaries, are home to elephant, bison, wolf, antelope, rabbit, fox, sambhar, wild boar, python, squirrel, blue bull, mongoose, jackal, honey badger, Malabar giant, tiger, deer, langur, rhesus, porcupine, wild cat, etc. Given the hot and arid climate, dry deciduous trees are common in the territory of Jharkhand. Madhya Pradesh with Omkareshwar temple has a subtropical climate with a sweltering and dry summer with temperatures ranging from 41 degrees Celsius to 25 degrees. The monsoon season runs from July and into September, with the southeastern parts of Madhya Pradesh area receiving 85 inches of rainfall annually and northwestern 40 inches. The area of Madhya Pradesh has been classified as being 61.7% reserved rainforest, 37.4% protected forest, and 0.9% unclassified forest. The total forest cover of Madhya Pradesh makes up 94.690 square kilometers. This means that approximately one-third of the state is covered by forest. Indigenous plants that grow there include teak, sal, bamboo, khair, and many medicinal plants, including, but not limited to, bael tree, *Aegle Marmelos*, an evergreen tropical shrub *Bixa orellana*, leguminous *Cassia tora*, succulent *Aloe barbadensis*, *Acorus calamus*, Palash *Butea monosperma*, Mango *Mangifera indica*, shrub *Woodfordia fruticosa*, and many more. The climate of Varanasi, where Vishwanath temple is located, is hot, with summers ranging from 92 to 110 F, winters cooling down to only 60–72 F, and 24+ inches of rain in monsoon. The state is rich in flora with 1015 species of plants with a mix of Fabaceae (legumes), Asteraceae (perennials), and Poaceae (grass). Nasik city, where Triambakeswhar temple is located, falls in the Western Ghats, with a hot and dry climate with highs of 111 degrees Fahrenheit and lows of the 50s. Nasik has large cultivable land, with 663,200 hectares growing kharif, monsoon crop; 136,500 hectares with rabi, winter crop, the forested land of 340,000 hectares (21.75%); and the uncultivable area is 23,000 hectares (1.48%) (Nasik district, 2020). About 5% of cultivable land grows medicinal plants, another 5% the horticulture (Lakshminarasimhan and Sharma 1991). Animal breeding is a robust economic activity as well. As per 2012 survey, Nasik accounted for 10,10,005 cows, 2,61,390 buffalo, 3,50,783 sheep, 5,33,068 goats, and 11,16,569 poultry (Nasik government, 2021). The data for wildlife is not accessible. Rameshwaram is amid the Indian oceanic strait. It is home to marine life, seaweeds, shells, fishes, and associated microbial life onshore and inland. The Sethu Samudram Canal in the Gulf of Mannar at the confluence of the Bay of Bengal, the Arabian Sea, and the Indian Ocean and its ocean floor has been declared a Marine Biosphere Reserve (MBR). The reserve is a highly productive area in the Indo-Pacific region that provides a rich habitat with a great biodiversity of 5896 species of plants and animals ranging from bacteria to dugongs (Kumaraguru et al. 2000; Kannan et al. 2001; Subba Rao et al. 2008).

In some cases, the landscape design within the prakara and parisar references the climate by selecting local trees to be planted within prakara and in proximate parisar. Still, it falls short in consciously paying homage to or playing a subtle educational role in communicating the region's biodiversity to the visitors.

Chapter 6
Reflection and Projection

All the Shiva temples are iconic, with a significant religious value attached to them, and thus an integral part of intersecting social, religious, spiritual, tourism, economic, and political ecosystems all nested within the complex, built environment, further nested in the local, regional, and bio-physical ecosystem.

Urbanism and Ecology

The ancient Shiva temples, considered significant for the Hindu pilgrimage, act as an effective anchor of the town, generating several economic micro and macro ecosystems around them. Other temples are added over time to these spiritual centers, around the anchor of ancient temples or yajna sites (Udayakumar 1995) to leverage the established ecosystems and infrastructure.

The layout of temple town Mandala varies with location and context. Baijnath, Triambakeshwar, and Vishwanath are in dense urban contexts, with varying built density, geography, and economics. Baijnath and Triambakeshwar are resonant with the Padmaka mandala, which allows for controlled containment and expansion. While Baijnath and Triambakeshwar have open landmarking the edges and some distance between the subsequent settlement, Vishwanath has a relatively more intensely built physical context contiguous with similarly intense built settlements/towns/cities.

Omkareshwar and Rameshwaram are both located on an island and yet starkly different as the former island is bound by a river abutting land on both sides. At the same time, the latter is an island in the middle of an ocean. Yet, both reflect a different Mandala, a simpler Dandaka for Omkareshwar and Padmaka (just with the southeast quarter missing) for Rameshwaram. The response projects a consideration of the diverse natural contexts of the two temples/towns. Kedarnath is similar to both Omkareshwar and Rameshwaram because they are nested in areas marked

A. Sharma, *Mandala Urbanism, Landscape, and Ecology*,
https://doi.org/10.1007/978-3-030-87285-4_6

with natural boundaries, with Rameshwar and Kedarnath contextualized in most extreme geographic and climatic conditions, of snowy mountains and a fierce ocean.

Philosophically and spiritually speaking, the temple perhaps does not want to be found by everybody but only the one who is elevated to the higher level of consciousness to withstand the journey from the mainland to the upland, metaphorically speaking, and has renounced the comforts to arrive here. Perhaps this is when one gets closer to becoming the image of a Sadhu. This can be considered a rather unusual view ranging from grim to romantic or stoic.

Simply speaking, do not build the settlements or provide any easy transportation between.

The Kedarnath temple is nested at the foothills of the steep Himalayan mountain terrain. The accommodations for the caretakers, tourists, and associated economic activities are located south of the temple. Given the topography, this is the natural path of the glaciers or mudslides. When the design and development decisions are constructed and placed on land, they become a part of the prevailing ecosystem. No matter how auspicious a mandala and layout form is, if placed in the path of nature to obstruct or work against it, it will only yield inauspicious results. Thus, even though most temple site and plot shapes, as shown in figure-ground abstractions, in appendices, confirm the auspicious shape, it will not prevent calamities from happening if sited in the wrong location, as with the case of the Kedarnath temple.

From the vantage point of religious belief, a temple is a symbolic embodiment of the energy representing the deity. A daily ritualistic service needs to be maintained to sustain and enhance the pranic energy of the place. This implies the requirement for priests and caretakers and infrastructure to meet their basic living needs. Revenue is required to maintain these services and the temple's upkeep. Therefore, revenue generation through tourism is an excellent supply to feed this existential economic ecosystem. Geo-politically speaking, the shrine creates an ecosystem of pilgrimage tourism between the Panch-Kedar sites, including Badrinath to the east of Kedarnath; these form the last non-militarily inhabited bastion for India, before the border crossing into China, thus forming a critical terrain to have control and eyes over.

Accepting that the settlements in Kedarnath are necessary, the settlement's design and layout still need to be reviewed. The decision is obvious regarding locations to be avoided, which is the natural path of glaciers and mudslides. However, theoretically speculating, the Dandaka layout would have been suitable for the current location of settlement, as it allows for the Welement flows to pass through the liner passage grids of the settlement. In view of current topography, potential mass movement, and flow paths, the settlement should be relocated to prevent future fatality. The land suitability analysis per rigors of geological and landscape systems analysis and studies of tectonic, sub-terraneous movements, processes, and other associated sciences needs to be undertaken for such a re-location recommendation. However, just based on the visual reconnaissance of a basic satellite imagery of the area, it seems evident that moving the settlement to the east of current location, on the stretch of terrain where Bhairavnath temple is located, will be the best move, shifting the settlement out of the direct flow path of the glaciers, until other studies are completed. When humans are preparing to land on Mars, Kedarnath re-settlement could be used as an opportunity to explore ingenuity with thinking on design, planning, and construction. Essentially, re-conceptualizing the architecture and siting of the settlements, imagined from the notion of co-existing with the "given" landscape, should be the core consideration, especially for rugged terrains.

Social congregation within the prakara are (a) along the processional path/line for darshan of the shrine, (b) to use basic amenities, (c) to rest after a tiring walk, (d) to buy offerings for the shrine, or (e) to buy prasad. The congregations within the parisar also follow the same patterns except that the vendors and amenities are all at compounded scale, quantity, and intensity and complemented with (f) souvenir shops and other retail, (h) restaurants, and (i) accommodation.

The congregation patterns within the prakara can primarily be addressed through design and layout and partially through regulation. Regulation on expanding the time of temple openings/darshan times to a wider window in peak seasons of temple visitation which are tied to climatic seasons would be helpful. The idea of temple Mandala if applied to a human body implies the human body as sanctum sanctorum or maha-mandapa and the space of ardha-mandapa as being the buffer space, if communicated to the visitors and devotees, to treat a recommended buffer space of 1 foot as sacred, non-penetrable space, which would bring a considerable order to social congregation practices within the temple and prakara, especially useful during the post-Covid times.

The parisar requires a combined intervention of design, planning, licensing of services, and regulations. However, the Mandala can offer cues to buffer zones for spacing activities concerning each other but also distancing them from the temple prakara, to maintain the sanctity of processional walk or journey to the shrine, which currently, oftentimes colludes in the chaos of commercial activities right outside the temple prakara/perimeter, thus jarring a contemplating soul into the layered existential realities of the physical, the commercial, and the spiritual all juxtaposed on another, in a single moment of arrival.

The policies on form-based code for architectural style, building heights, and selection of materials and colors, which can be used for built environment in the temple parisar, to set off the ancient grandeur of the temple, can be employed to create a more spatially organized and experientially peaceful experience of these temple towns. At the same time, policies preventing the use of contemporary trappings and materials for temple upkeep and prakara expansion should be put in place to bring some semblance of periodic times of the temples.

Landscape and Ecology

The current landscape within the temple prakara looks like an after-thought and a contrived pattern to fit within a randomly expanding temple perimeter. If reconciled with traditional Mandalas, this practice may yield more aesthetically pleasing landscape form with options of a rich variety of spatial and landscape organization.

The study of temple town presented in the earlier chapter shows a significant imbalance in the built and natural environments. Reintroducing the classic principles from Arthashastra and Manasara, as shared in the second chapter, such as systemic and incrementally expanding green areas and forested belts planted at the outer perimeter, would balance the natural ecology with that of the built environment. The towns and biodiversity could use the three-tiered green belt as a breathing space for outdoor gazing at or contemplating in nature, passive recreation/pleasure gardens and medicinal/Ayurvedic plants, and tree-based agroforestry. Planting of the forested belts, as proposed by Mandala (Mookerji, 1960) for around the town, should be considered for plantation around the temple. This will be in sync with the initial founding of the shrine, which was mostly in forested or other natural landscape, with the forest and tree groves around them considered sacred. Sacred groves should be planted and tended around the temple. This can be integrated within the economic ecosystem of the temples as well as an option of archana for tree grove or tree planting. Also, the monetary donations given to the temple, which are used toward operational expenses and social causes, could be diverted in part to the planting of sacred tree groves, including the dedicated groves of medicinal plants.

Water features need to be reconsidered as a part of the natural ecosystem and redesigned with preservation and an active functional interface in mind. Expanding the design in such a way that the natural spring of water body remains untouched and other water features are planned around it with a diversion of minor water streams from the holy Kundas to other water features, which not only act as active ablution ponds for devotees but also perform the function of collecting rainwater. Stormwater management could be ingeniously developed and designed as such and as running through the entire layout of the town as well. Similarly, the water tanks should be multiplied by creating satellite water features to the original, organized as a network of water Kundas, with only one or few of those open for ritualistic ablution and the rest operating as water harvesting or collection ponds for supply to the farms and town residents via a municipally controlled central water reservoir, when needed. The Mandala forms offer enough variations of size, shape, and formal partitioning to be deployed for installation of interim water harvesting and conservation Kundas through the town, stormwater diversion and collection, and active and passive recreation-based garden design in the city, besides the three-tiered forestry belt on the perimeter.

The puhspavatika/flowering gardens and latavaatikas/creeper trellised gardens being integrated into current landscape designs do not correspond with the spiritual ethos of the temple, as a multiplicity of fragrance and colors are associated with pleasure and thus distraction from the spirit of renunciation of the worldly to be one with the One/God. While it is not necessary to have a monotonous context to be able to meditate and in pursuit of finding God or simply for a brief communion with the Supreme Power/Param-eshwara, it offers a facilitative context and ambience of relatively higher stillness. Selection of planting materials is systematically sorted if guided by the religious beliefs and texts on most favorite and thus must-have and must-avoid listing of flowers and fruits corresponding with the deity.

Planting the forested belts around the town, as proposed in classic texts (Atharva shastra in Mookerji, 1960), should be revived, especially for densely built temple towns. The native trees and vegetation acclimatized to growing in the local climate should be used for these forestry and agroforestry belt plantations. The revival of increasing green cover using Mandala-based design and planning by Kamikagama, Manasara, and Mayamatam acts as a multi-faceted tool thus to revive ancient traditions in service of current needs of balancing the urban heat island effect created by the densely built environments and high human population, further aggravated by floating tourist population to the temple towns.

The Projective Landscape Typology

A mandala can offer a template for organizing urban landscape elements such as greenways (and greenbelts), green streets (large and small trees), urban agriculture, water collection ponds, pocket parks, and gardens in the city to serve the needs of climate change and urban sustainability. Some creative interpretations show the potential design reactions and thinking that even a most simplistic, geometric version of Mandala can spark in the urban landscape organization framework arena.

Here are a few graphic illustrations of environments possible through Mandala urbanism.

The Mandala was approached from a landscape architect or green urban planner's perspective and creatively rendered with landscape typologies of urban agricultural parks, gardens, water harvesting, filtering, cleaning best management practices, agroforestry green belts, and tree groves, as shown in the figure below. Different colors were attached to a different typology. However, as an experimental study, the colors do not matter. We are quickly reminded that the landscape primarily made of natural plants for green cover or urban agriculture soil and water are typically approached as fluid entities either fitting into a form or forming one. Different clustering combinations and siting are determined by the placement of the other structures within the geometry or place within the ecosystem of humans and human settlements (Fig. 6.1).

An illustrative exercise was conducted to theoretically speculate on the replicability of this concept from urban landscape architecture and design perspective. The mandalas with creative re-programming of the landscape typologies were applied to current temple towns to understand the spatial scope of an urban retrofit (Fig. 6.2).

A mandala can offer a template for organizing urban landscape elements such as greenways (and greenbelts), green streets (large and small trees), urban agriculture, water collection ponds, pocket parks, and gardens in the city to serve the needs of climate change and urban sustainability. Some creative interpretations show the potential design reactions and thinking that even a most simplistic, geometric version of Mandala can spark in the arena of urban landscape organization framework. Triambakeshwar and Rameshwaram, when viewed at a zoomed-out scale, presented the opportunity for designers to read compounded possibilities with the Mandala in

Fig. 6.1 Creative rendering of landscape typologies to Mandalas; illustration by Varun Gupta

terms of poly-nuclei radial. However, the landscape patterns and typology chosen for more Mahakaal, Somnath, and Vishvanath temples situated in relatively higher density urbanized settings, the designers chose quadrangular grids that meld well with current urban grids while allowing scope for marking focal quadrangles through programming and design. Landscape typology chosen for these dense urban settings mainly was related to urban greening and water features; it is possible that complex typology and details were left to be envisioned at a zoomed-in micro-site scale.

Fig. 6.2 Creative iteration of reprogramming Mandala with contemporary landscape typologies for temple towns: Triambakeshwar (top L and R), Rameshwaram (top R), Somnath (middle L), Mahaakaal (middle R), Kashi Vishwanath (bottom L), Rameshwar (bottom R), illustrations by Boghaf A. and Gupta V

Mandala Around the World

Globally, city plans have been discussed in terms of layout patterns, primarily grid or radial planning, in Western discourse for decades now. The discourse is rarely foregrounded in the mosaic of broader antecedent literature and frameworks occurring in the Asian context, such as the Vaastu Shastra in India.

Most post-industrial town planning and development in various cultures and geographies has been guided by transportation planning, primarily drawn for supplying missing connections or speed of travel between multiple points. The Mandala overlay exercise is not undertaken to claim the number of towns and city plans spread worldwide is based on Mandalas, as this is subject to history and anthropological research, but to read the projected congruence or dissonance. The exploration aids in the reflection on the usefulness of the Mandala urbanism to other towns than temple towns and different cultural geographies outside India. Even within India, the exercise of decoding Mandalas for temple towns sheds light on possibilities for domestic and national design thinking and planning agencies beyond the city of Jaipur (Chakrabarti 2013; Sinha 1998). For the purports of creative visioning using classic traditional knowledge for the future, within the scope of this book, it is a moot point of uncovering which cities were designed based on Mandala, but can we create better cities using Mandala formwork and how can we find solutions to the problems emerging today, using the same.

The figure below presents a reading of some international city grids corresponding to the classic Indian Mandala. Nandyavrutt Sarvatobhadra applies both to Washington D.C. in the United States and Barcelona in Spain. Barcelona has a land area of 39.2 square miles of land with a population of 1,620,343 and a density of 41,000 people per square mile. At the same time, Washington D.C. is roughly 1.5 times larger in the land area spread with 61.05 sq. mi of land area and one-quarter of the population density than Barcelona at 9856.5 people per square mile. This corroborates the physical geographic scale stretchability of the Mandala (Fig. 6.3).

Similarly, the Sarvatobhadra mandala can be read in the layouts of Barcelona in Spain and Brasilia in Brazil, with reflections of Dandaka in Brasilia and Sarvatobhadra and Prastara in Washington D.C. as well. Brasilia and Washington D.C. are both political capitals, with the latter being deemed more successful in spatial layout than the former, despite the continuing criticism. Amman, considering Jordan's political, cultural, and economic capital, also has a combination of Sarvatobhadra, Dandaka, and semi-radial Karmuka, showing up in different town quarters, each form corresponding with a unique functional purpose.

Fig. 6.3 Mandala around the world: Washington D.C. Nandyavrutt Mandala (top L); Barcelona, mosaic of Nandyavrutt and Sarvatobhadra Mandala (top R); Brasilia, Sarvatobhadra and Dandaka Mandala (middle L and R); Amman, Karmuka and Dandaka (bottom L and R, illustrations by Boghaf A. and Gupta V)

Chapter 7
Mandala Urbanism and Questions for Further Research

Sompura, who was the architect for many significant temples in India, observed that "the old codes of structural rules have their use today not so much in the construction of temples as in helping towards an understanding of the forms of the temples in the actual idiom of those who used to build them in the grand old days of the medieval era." (Sompura 1975, pgs 47,48). The thought has been extended to the town planning level in this book. To the critic, who constantly laments the presentation of any Hindu viewpoint as a threat to the rest, it can be revealed that the principles of temple and town planning are accessible to most towns with multiple faiths and applicable with modifications. The proposition of modifying and revising the mandala framework offers even more possibilities of inclusion. With the sophistication of collated thinking in terms of mathematics, religion, nature, and philosophy, with attention to universal social-cultural values that make a built environment hospitable, the mandala frameworks withhold the three-dimensional ideology to withhold the diversity or parts in its fold, while retaining the unified as a whole.

Mandala Urbanism Approach and Principles

City plans have been discussed in terms of layout patterns, primarily in terms of grid or radial planning. The narratives and critical expositions predominantly refer to European literature and the era of planning. The discourse thus presents a part of thinking, which needs to be foregrounded in the mosaic of broader antecedent literature and design and planning thinking occurring in the Asian context. This book sheds light on the grid planning discourse in India and through Vedic times.

Different periods presented different concerns to be addressed through differently unfolding ecologies of social-environmental co-existence, with economics either guiding or serving the parts and interconnections as prioritized by society at the time. Although being discussed with paramount urgency now, climate change is

A. Sharma, *Mandala Urbanism, Landscape, and Ecology*, https://doi.org/10.1007/978-3-030-87285-4_7

not a phenomenon of the recent decade but has been at play since the initial scientific recording of the planetary systems. Thus, we read about the ice age and the role of carbon and methane—the greenhouse gases—in warming up the climate, enough to facilitate inception, growth, and sustenance of human life. Human society's desire to sustain the human species now presents a situation that demands a resolution to the greenhouse gases to stop the warming to prolong the life and health of a natural ecosystem, critical for the survival of human life, but which develops an industrial system (IS) that disrupts the natural ecosystem.

There are some postulated that should explain the importance of natural ecosystem (NES):

- If there is no natural ecosystem, there may not be any human life.
- If there is no human life, there would still be a natural ecosystem.
- If there is no abiotic system, but there is a natural ecosystem, then human life could still exist (with a different level of comfort and lifestyle).

The different versions of urbanisms that have been explored over time, such as new urbanism, transport-oriented development, smart growth, green urbanism, and sustainable urbanism, all ascribe to the ideology of compact walkable, bikeable development, mixed land use, efficient transportation system, social-recreational spaces, and vibrant, livable communities (Farr 2011, 2007, Duany and Plater-Zyberk 2006a, b, 2021). Distinctive planning principles include the focus on prevention of urban sprawl and connections between housing and business under tenets of *new* urbanism. Transport-oriented development emphasizes development around train stations, or to expand it further, around public transit systems, as a strategy of suburban renewal and creation of sustainable communities. Smart growth starts adding considerations regarding range of housing options; distinctive neighborhoods with a strong sense of place; variety of transportation choices; inclusive decision-making with community and stakeholders through the development process; making development predictable, fair, and cost-effective; as well as preservation of open spaces. Sustainable urbanism adds to the conversation through recommendations on reconsidered focus on neighborhoods, decentralized governance and reinforcement of city's identity, long-term vision for a place, design quality and fairness and trust-building with the community, and partnerships incentives for economic investments. The range of urbanisms discussed above considers natural ecosystems the other systems and emphasizes designing the built industrial system that exerts low impact on the natural ecosystem. Several approaches take the intermediary approach. Climate-centered urban planning is advocated by some, emphasizing increased attention to climatic factors such as humidity, air, temperature, radiation, winds, and clouds (Luca 2017). Sustainable green urbanism advocates for minimizing the use of energy, water, and materials at each life cycle stage of a city, including the embodied energy in the extraction and transportation of materials, their fabrication, their assembly into the buildings, and, ultimately, the ease and value of their recycling when an individual building's life is over (Farr 2011, 2007).

The biophilic urbanism approach brings the built environment focus of green urbanism to open landscapes and champions for an increase in urban gardens and natural resources to make the city more accessible, natural, and beautiful but also to mitigate the harsh environmental impacts of climate change such as rising temperatures (Lee and Youngchul 2021). While tactical urbanism tests the approach of shaping social, ecological, and economics change in a place around a formal insertion, sponge urbanism (Sowell and Weidemann 2009) and water-sensitive urbanism (Rising 2015, Schneider, David and Andrew 1973) are amenable to be formed by the formal, structural guide. It is worth exploring if Mandala can respond to this opportunity and challenge. Mandala urbanism follows the landscape characteristics with all its peculiarities, allows for a considerable heterogeneity of spatial volumes, and additionally considers the natural ecosystem as a part of the ecosystem at every incremental step of design, planning, and organizational thinking, thus making a consistent effort to be in harmony with the natural ecosystem at every nested hierarchy scale. Mandala urbanism thus addresses the critique of grid planning for its role in destroying the pre-development landscape and its peculiarities (Ben-Joseph and Gordon 2000) and re-presents a possibility.

The Mandala urbanism approach:

- Does not necessarily warrant the annihilation of the built *industrial system* (IS) but offers grids as a guide to integrate, distribute, and yet contain the abiotic industrial system.
- Is premised in divining the *industrial system* (IS) for best fit with *the natural ecosystem (NES).* NES or climate is not contained or controlled by grids, but all-pervasive and not bound by grids, but grids offer a control mechanism and sub-unit to create and maintain balance the two systems.
- Calls for an iterative check on the neighborhoods and towns to assess the balance of IS with NES. The concept of ecological carrying capacity can be used to develop an even more robust model of assessment to gauge the balance of the two systems, within itself (neighborhood) and across systems (network of neighborhood = town) and then across Mandalas (network of towns/cities= region).

Mandala urbanism takes a view on balancing all the five natural elements (panchmahabhoot) and matter (dhaatu), with cognizance of all postulates, at every nested hierarchy scale of a town city or planet. It ascribes to the design syntax approach of decoding and recoding the city that offers a syntactic tool to re-assembling, re-programming, and re-crafting the spatial sub-units of the city.

Principles of mandala urbanism:

- Functional goal-oriented: defense, leisure, education, commerce, spiritual, holistic
- Geographical and socio-cultural context-responsive
- Climate-sensitive, climate-influenced, working with climate
  - Balancing the built industrial system (IS) in harmony with natural ecosystem (NES) at every nested hierarchy scale of a town, city, or planet
- Spatial variations:
  - Core, periphery, and grid sections proportions: proportional hierarchy

  - Complex combinatorial grids of varying proportions disrupting the idea of core, periphery, and grid
    - Consideration of the built, industrial system (IS) and natural ecosystem (NES) as syntactic tools to decode and recode, re-assemble, re-program, and re-craft the spatial sub-units of the city toward a harmonized Mandala and city.
    - Does the programming of sub-units allow for more control in creating inclusive neighborhoods?
- Concept of Pradakshina path as a guide to the movement and flows, constantly moving from periphery to center; linear passages only through periphery; and a quick passage from the periphery to and through the center, resulting in abrupt arrivals or surprise moments
  - Does the pradakshina path offer a way to organize infrastructure for (climate-manmade) disaster-safe urbanism?
  - Can the pradakshina path and attached spatial grid units allow a systematic space for sponge urbanism, agroecological urbanism, green infrastructure, and urbanism?
  - A mandala can be coded to ensure that locational decisions are weighted against neighborhoods of lower economic means but allow for siting of environmentally renewable industries and facilities to render maximum benefits and minimum negative impact for all communities and all biodiversity of plants, birds, and animals?
- Concept of Marmasthana—fragile areas to be preserved
  - Placement of industrial system (IS) in harmony with natural ecosystem (NES)
- Strong emphasis on symmetry underscores the concept of balance
- Pradakshina path as potential internet of grids with the built environment and regenerative landscapes

New Compound Mandala

The combinations and permutations of Mandalas and sub-units within are as varied as mathematically arrived through N=NPR and N=NCR but even more when creatively derived, through repetition. Table 7.1 presents the mathematical denomination of Mandalas as conceived by the author to derive permutation and combination algorithms.

Table 7.1 The mathematical denominations for Mandala

	Quadrangular organization form work		
1	Prastara	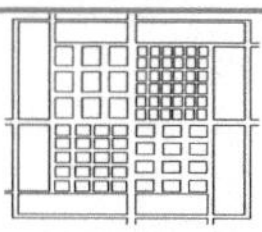	Prastara square = PRSq
		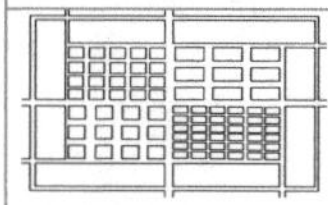	Prastara rectangle = PRrec
2	Nandyavrut	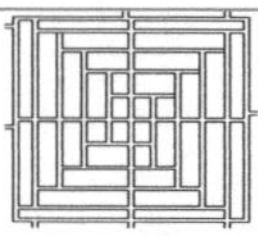	Nandyavrut square = NVsq
		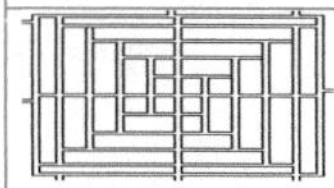	Nandyavrut rectangle = NVrec
3	Dandaka	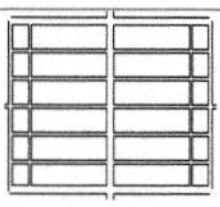	Dandaka square = DAsq
		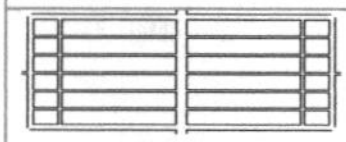	Dandaka rectangle = DArec
4	Sarvatobhadra	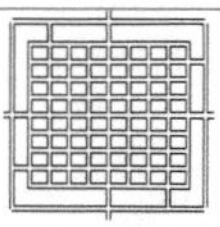	Sarvatobhadra square = SBsq
		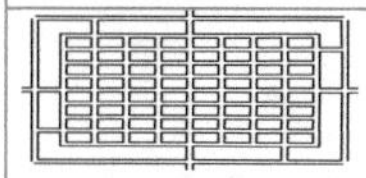	Sarvatobhadra rectangle = SBrec
5	Rajadhaniya	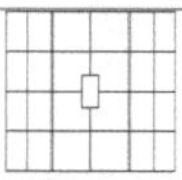	Rajadhaniya square = RDsq
		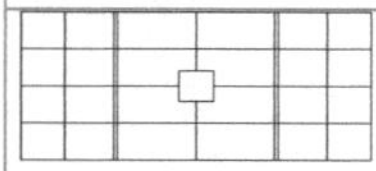	Rajadhaniya rectangle = RDrec
6	Chaturmukha	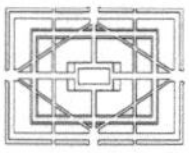	Chaturmukha square = CMsq
		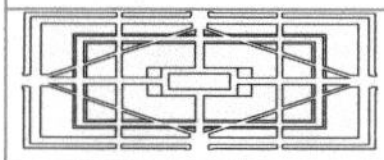	Chaturmukha rectangle = CMrec
7	Khetaka	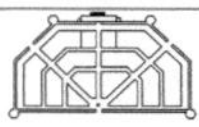	Khetaka polygonal = KHpoly
8	Kubjaka	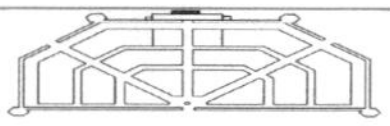	Kubjaka polygonal = KUpoly

(continued)

Table 7.1 (continued)

	Quadrangular organization form work		
Radial organization formwork			
9	Nandyavrutt		Nandyavrutt Circular = NVcir
			Nandyavrutt Oblong = NVob
10	Padmaka		Padmaka Circular = PAcir
			Padmaka Oblong = PArec
11	Knarvata		Knarvata Circular = KNcir
			Knarvata Oblong = KNob
12	Karmuka		Karmuka semicircular = KAscir

Source: Table prepared by author, the mathematical denominations to the Mandalas have been attributed by the author

Mathematical Denomination of Mandalas

So, there are 21 numbers of Mandala variations as annotated in the table below.

1. PRSq
2. PRrec
3. NVsq
4. NVrec
5. DAsq
6. DArec
7. SBsq
8. SBrec
9. RAsq
10. RArec
11. CMsq
12. CMrec
13. KHpoly
14. KUpoly

15. NVcir
16. NVob
17. PAcir
18. PAob
19. KNcir
20. KNob
21. KAscir

Algorithms for Mandala Permutations and Combinations, Developed by the Author

Algorithms for potential formations
∑PLAN (City/ Town/ Neighborhood/ Building / Landscape/ Site plan)= 1PRsq +1(PRSq/ PRSq/ PRrec/ NVsq/ NVrec/ DAsq/DArec/ SBsq/ SBrec/ RAsq/ RArec/ CMsq/ CMrec/ KHpoly/ KUpoly/ NVcir/ NVob/ PAcir/ PAob/ KNcir/ KNob/ KAscir), Where ∑ refers to Summation, PLAN refers to an organization pattern, physical, architectural, engineering, or site plan for the scales of a City/ Town/ Neighborhood/ Building / Landscape/ Site/ Planting plan, or any part thereof, but also applicable to any product or part thereof, hand drawn or digital), and other abbreviations correspond to Table: Mathematical denomination of Mandalas, or any derivative, as is or with content or conceptual similarity)
∑ PLAN = 1PRsq +1/2 (PRSq/ PRSq/ PRrec/ NVsq/ NVrec/ DAsq/DArec/ SBsq/ SBrec/RAsq/ RArec/ CMsq/ CMrec/ KHpoly/ KUpoly/ NVcir/ NVob/ PAcir/ PAob/ KNcir/ KNob/ KAscir) Or for purpose of generic abstraction, let us call remove the Mandala specificities, and call them as M1, M2 mandalas or patterns. ∑ PLAN (MandalaTotal) = "x" fraction of M1 + "y" fraction of M2, Where center of M1 could be used as the 0,0 base coordinate, in x,y coordinates in quadrangular coordinate system for a 2 dimensional spatial organization, and at 0,0,0, in x,y,z in cylindrical coordinate system.
∑ PLAN = 1PRsq +1/4 (PRSq/ PRSq/ PRrec/ NVsq/ NVrec/ DAsq/DArec/ SBsq/ SBrec/RAsq/ RArec/ CMsq/ CMrec/ KHpoly/ KUpoly/ NVcir/ NVob/ PAcir/ PAob/ KNcir/ KNob/ KAscir)
∑ PLAN = 1PRsq +1/8 (PRSq/ PRSq/ PRrec/ NVsq/ NVrec/ DAsq/DArec/ SBsq/ SBrec/RAsq/ RArec/ CMsq/ CMrec/ KHpoly/ KUpoly/ NVcir/ NVob/ PAcir/ PAob/ KNcir/ KNob/ KAscir)
∑ PLAN = 1PRsq +1/12 or any geomatrically and mathematically possible fraction of (PRSq/ PRSq/ PRrec/ NVsq/ NVrec/ DAsq/DArec/ SBsq/ SBrec/RAsq/ RArec/ CMsq/ CMrec/ KHpoly/ KUpoly/ NVcir/ NVob/ PAcir/ PAob/ KNcir/ KNob/ KAscir)
∑ PLAN =1 (or any geometrically and mathematically possible fraction of) PRsq + 1 (or any geometrically and mathematically possible fraction of) (PRSq/ PRSq/ PRrec/ NVsq/ NVrec/ DAsq/DArec/ SBsq/ SBrec/RAsq/ RArec/ CMsq/ CMrec/ KHpoly/ KUpoly/ NVcir/ NVob/ PAcir/ PAob/ KNcir/ KNob/ KAscir)

EXAMPLE

For example,
∑ PLAN = "x" fraction of PAcir + "y" fraction of PRrec, no conditions attached,
Or for purpose of generic abstraction, let us call remove the Mandala specificities, and call them as M1, M2 mandalas or patterns.
∑ PLAN (MandalaTotal) = "x" fraction of M1 + "y" fraction of M2,
Where center of M1 could be used as the 0,0 base coordinate, in x,y coordinates in quadrangular coordinate system for a 2 dimensional spatial organization, and at 0,0,0, in x,y,z in cylindrical coordinate system.
∑ PLAN Khand **Prastarit Padmaka** = "x" fraction of PAcir + "y" fraction of PRrec,

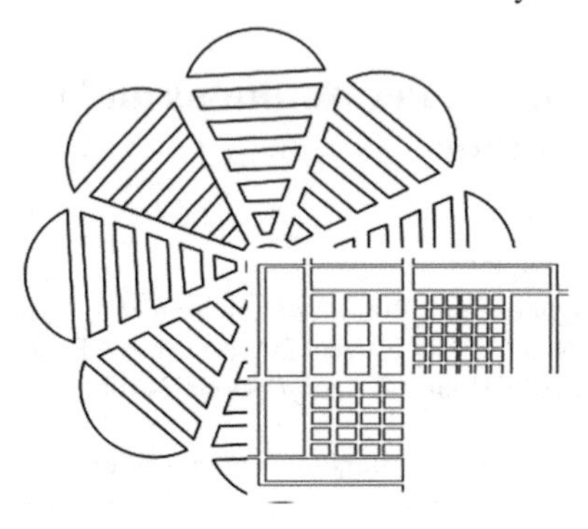

∑ PLAN = "x" fraction of PAcir + "y" fraction of PRrec, conditions attached
Condition 1:
∑ PLAN (**PrastaraPakshaDandaka**) = 1/2 PAcir + 1/2 PRrec, where the PAcir and PArec equidistant central/mid-way point of one edge of PAcir aligns/converges, facing the equidistant central/mid-way point of one edge of PArec

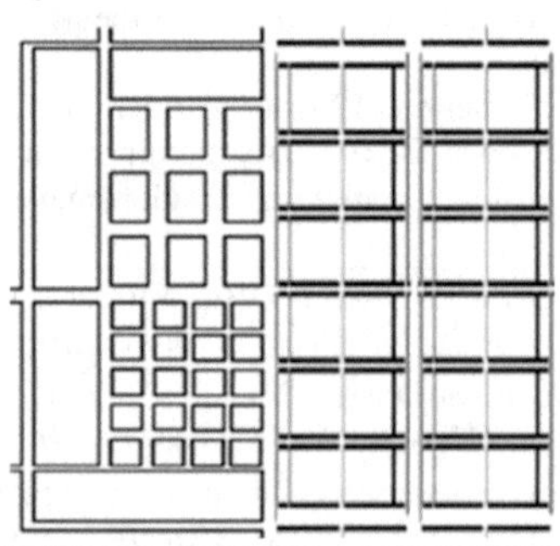

∑ PLAN (**Prastarit Padmaka**) = "1" fraction of PAcir + "1" fraction of PRrec, however, 1PAcir+ 1PRrec,
condition 2: where 1PRrec, can be smaller than PAcir by any percentage, but the centre of both PAcir and PRrec the form must match

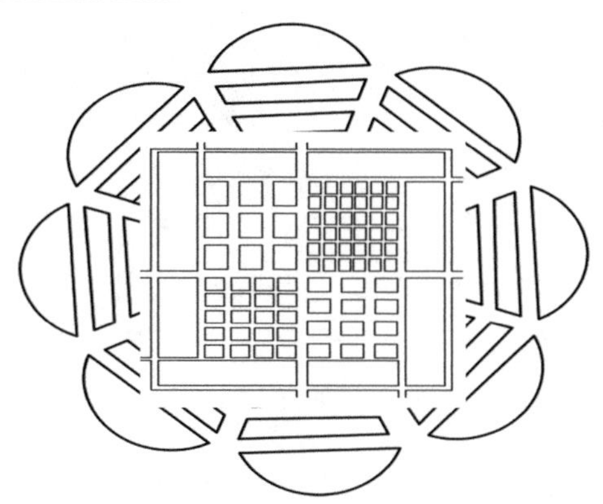

∑ PLAN Prastarit Padmaka = "1" fraction of PAcir + "1" fraction of PRrec, however, 1PAcir+ 1PRrec,
condition 3: where 1PRrec, can be smaller than PAcir by any percentage, and the centre of PArec can shift anywhere on the background mosaic of PAcir

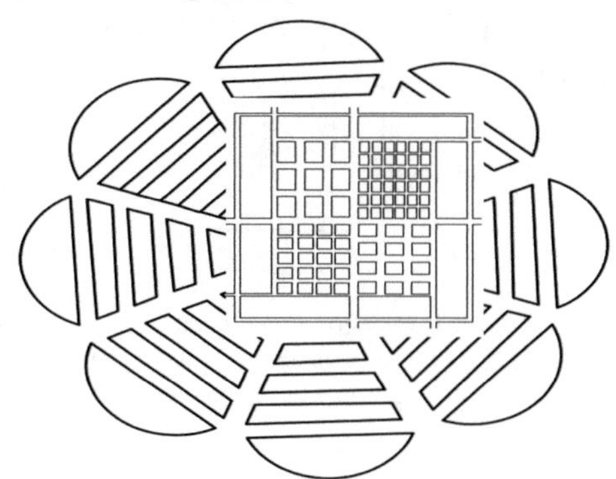

Another example of this equation and condition, is
∑ PLAN **Dandamandit Sarvatobhadra**

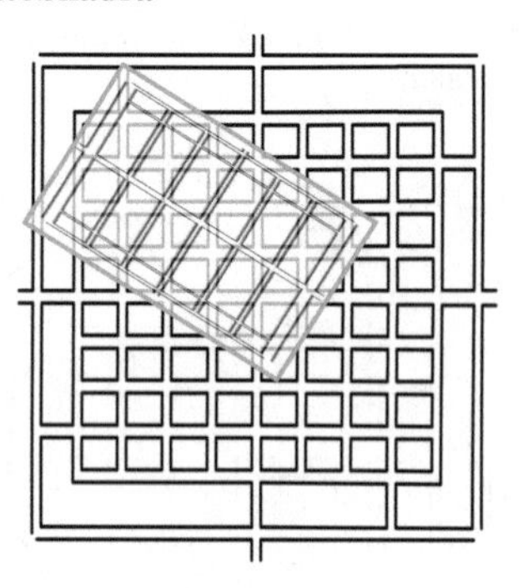

∑ PLAN (**Mandala Mandal**):: "x" fraction of M1 + M2+ (any number of Mandala forms), conditions attached,

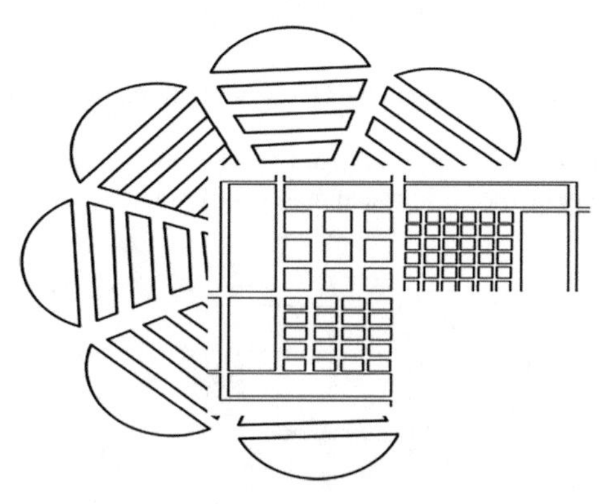

Condition 4: where M or ny fractional variation of M, is co-located in a region, with a distance between the center but also edges, but with at least one, line, points or any geometrically possible physical, connection between either centers, edges, or any point within the M,

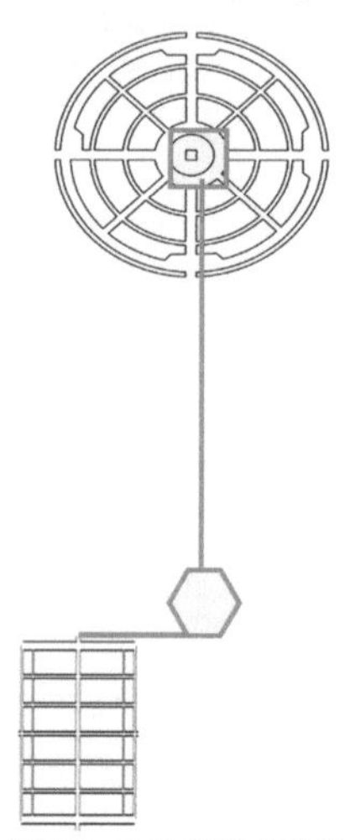

$\sum$ Form (**3D Mandala), F**: "x", a fraction/ value of PAcir/PRrec/ any number of Mandala forms + "y", a fraction/ value of PAcir/PRrec/ any number of Mandala forms + "Z", a fraction/ value of PAcir/PRrec/ any number of Mandala forms

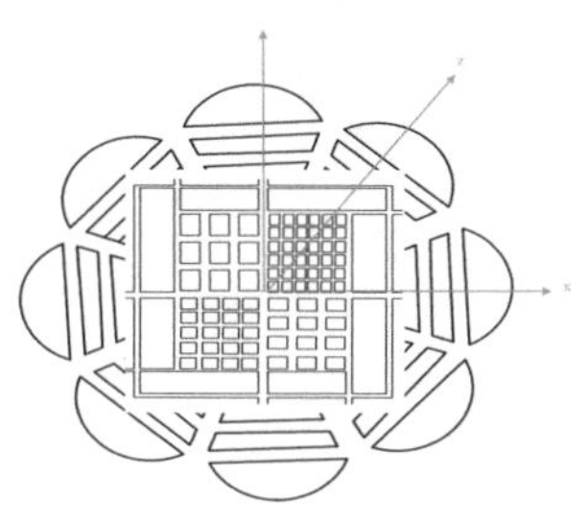

Mandala urbanism can be employed to structure rapidly increasing urbanization. Mandala urbanism can be used to control urban sprawl, reinvigorate suburban development, and create healthy, vibrant neighborhoods. The new compound Mandala offers multi-dimensional spatial organizational possibilities, through permutation combination, with equally diverse programming options to create communities that flourish and ecologies that self-propagate and regenerate, with a healthy balance and "niche" for all species.

Here are some questions for further deliberation:

- Is it more effective to apply this approach at every scale and hierarchy of human interventions in the ecosystem? And consequently, use the IS and NES balance as a Mandala urbanism (MU) assessment framework to assess the total ecosystem health of the town, city, country, and planet at each nested hierarchy?
- Can three-dimensional extrapolations of the mandala be used as a formwork to re-code the city's spatial, socio-cultural, ecological, and economic intranet?
- How does it translate into policy? Does a neighborhood or town that generates enough oxygen through a natural ecosystem, with close to zero negative impact, get O2 oxygen tax incentive credits to support its human population? Should these credits be non-tradable and non-transferable among neighborhoods or any geopolitical settlement designation except for cases that might be suffering from natural resources constraints such as lack of land, water, and air to nourish plant trees and vegetation to generate the oxygen?

Irrespective of the type of urbanism, sociology-centered calls for significant monetary investments into poorer communities to transform them from crime-ridden unsafe places into investment-friendly, safe zones, which will have to be attended, through policies.

More than Mars-explorers, we need earth-keepers and sustainable earth urbanism that the Mandala urbanism attempts to, iteratively, create.

Appendices

Appendix I. List of Site Planning Mandalas Showing Number of Grid Components (Acharya 1927a, b, 1934a, b, c)

	Name of the site planning mandala	Number of parts of the grids		Name of the site planning mandala	Number of parts of the grids
1	Sakala	1	17	Triyukta	289
2	Pechaka	4	18	Karnashtak	324
3	Peeth	9	19	Ganit	369
4	Mahapeeth	16	20	Suryavishalak	400
5	Upapeeth	25	21	Susahit	441
6	Ugrapeeth	36	22	Supratikant	484
7	Sthandil	49	23	Vishaalak	529
8	Chandil or Mandook	64	24	Vipragarbha	576
9	Param Shaayik	81	25	Vishvesh	635
10	A(us)san	100	26	Vipulbhog	676
11	Sthaaneeya	124	27	Viprakant	729
12	Deshya	144	28	Vishaalaaksha	784
13	Ubhaychandit	169	29	Viprabhakti	841
14	Bhadra	196	30	Vishveshwar	900
15	Mahaasan	224	31	Ishwarkaant	961
16	Padmagarbha	256	32	Indrakaant	1024

A. Sharma, *Mandala Urbanism, Landscape, and Ecology*,
https://doi.org/10.1007/978-3-030-87285-4

Appendix II. Site Planning and Town Layout Mandalas (Acharya 1927a, b, 1934a, b, c)

Site plan: Sakala—1 plot

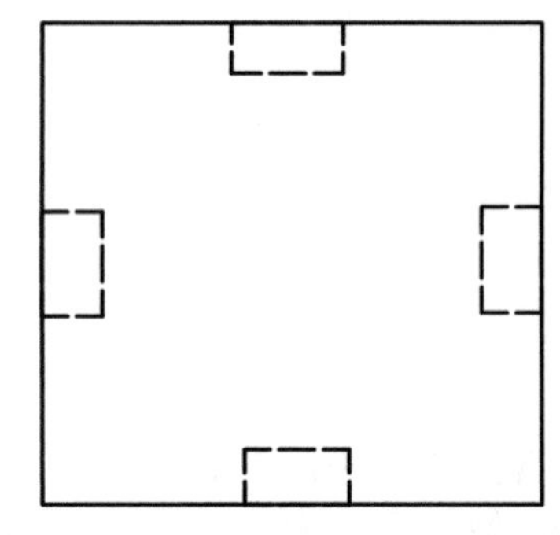

Site plan: Pechaka—4 plots

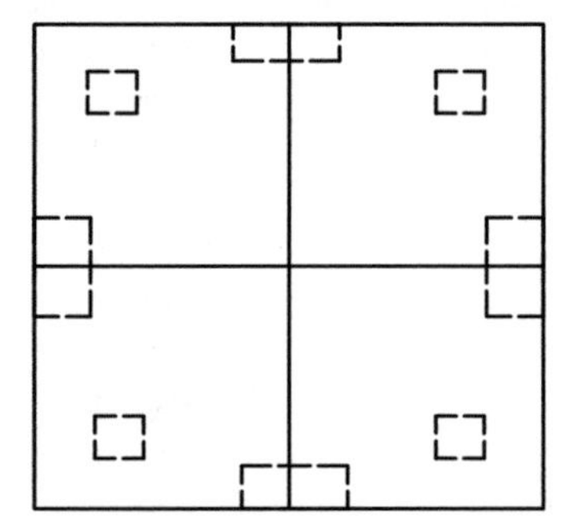

Site plan: Pitha—9 plots

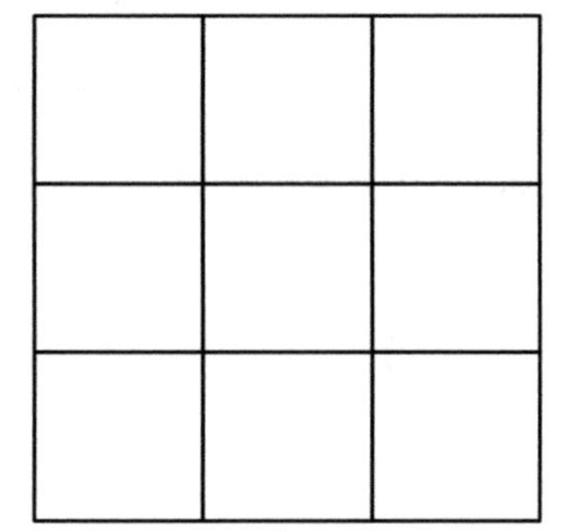

Site plan: Mahapitha—16 plots, arrangement 1

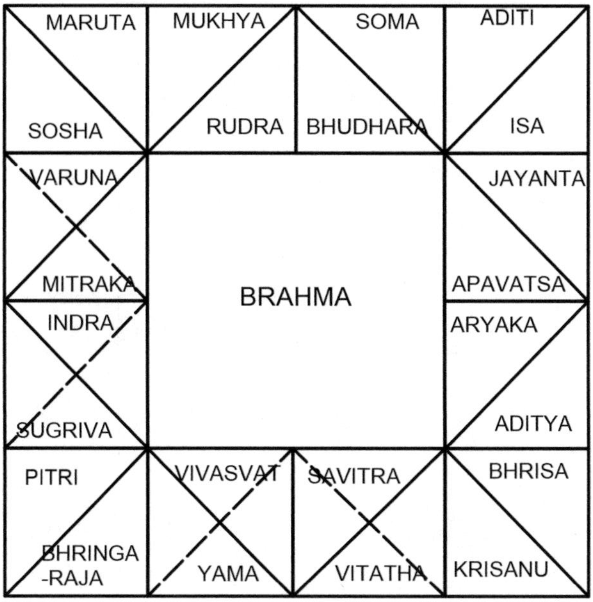

Site plan: Mahapitha—16 plot, arrangement 2

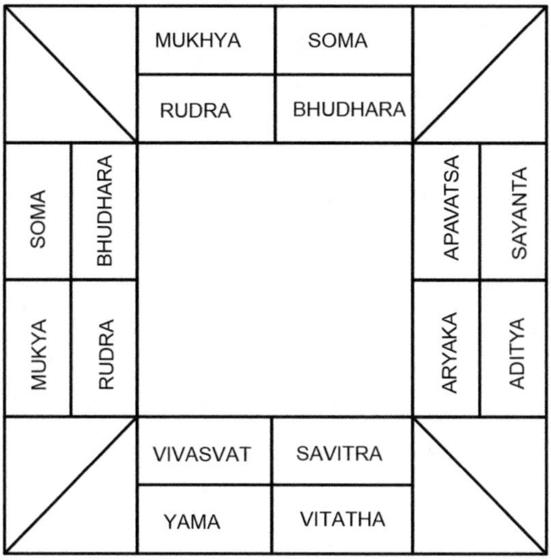

Site plan: Upapitha plan—25 plots

MARUT	MUKHYA	SOMA	ADITI	ISA
SOSHA	RUDRA	BHUDHARA	APAVATSA	JAYANTA
VARUNA	MITRA	BRAHMA	ARYAKA	ADITYA
SUGRIVA	INDRA	VIVASVAT	SAVITRA	BHRISA
PITRA	BHRINGA -RAJA	YAMA	VITATHA	AGNI

The Ugra-pitha plan—36 plots

VAYU	MUKHYA	SOMA	ADITI	ISA
SOSHA	RUDRA	BHUDHARA	APAVATSA	JAYANTA
VARUNA	MITRA	BRAHMA	ARYAKA	ADITYA
SUGRIVA	INDRA	VIVASVAT	SAVITRA	BHRISA
PITRA	BHRINGA -RAJA	YAMA	VITATHA	AGNI

Site plan: Sthandila plan—49 plots

VAYU	MUKHYA	SOMA	ADITI	ISA
SOSHA	RUDRA	BHUDHARA	APAVATSA	JAYANTA
VARUNA	MITRA	BRAHMA	ARYAKA	ADITYA
SUGRIVA	INDRA	VIVASVAT	SAVITRA	BHRISA
PITRA	BHRINGA-RAJA	YAMA	VITATHA	AGNI

Site plan: Chandita plan—64 plots, quadrangular

NAGA / VAYU	UDITA	BHALLATA	SOMA	BHRINGA-RAJA	ADITI	JAYANTA	ISANA / PARJAHYA
MUKHYA	RUDRA-JAYA / RUDRA					APAVATSA / APAVATSA	ANTA-RIKGHA
ROGA		MITRA	BHUDHARA		ARYAKA	MAHENDRA	
SOSHA			BRAHMA			DINAKA	
VARUNA						SATYA	
PUSHPA-DANTA			VIVASVAT			BHRISA	
GODHA	INDRA-RAJA / INDRA	MRISA	GANDHARVA	YAMA	RAKSHASA	SAVITRA / SAVITRA	VITATHA
DAUVARIKA / PITRA	SUGRIVA					MRIGA	AGNI / PUSHAN

Site plan: Chandita plan—64 plots, radial

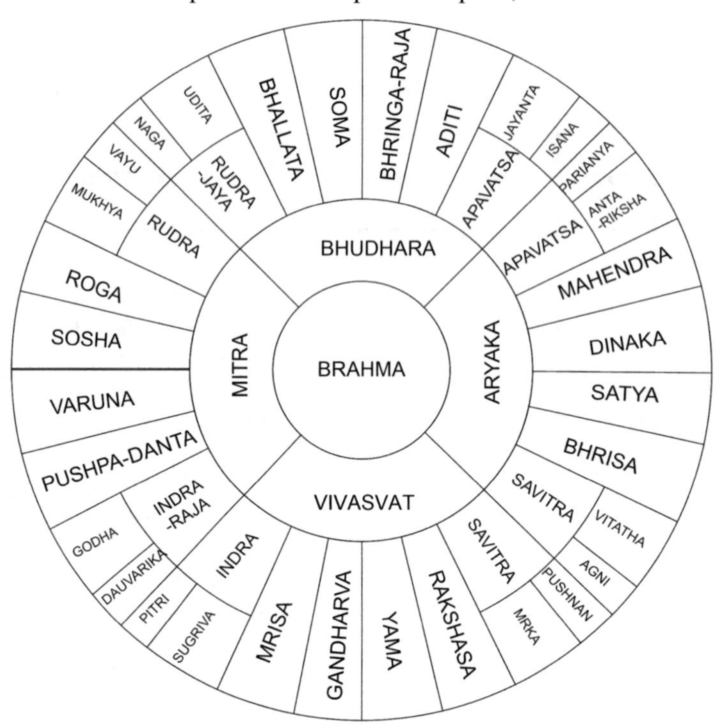

Site plan: Param sayika plan—81 plots, quadrangular

<table>
<tr><td>MARUT</td><td>NAGA</td><td>MUKHYA</td><td>BHALLATA</td><td>SOMA</td><td>MRIGA</td><td>ADITI</td><td>UDITA</td><td>ISA</td></tr>
<tr><td>ROGA</td><td rowspan="2">RUDRA</td><td rowspan="2">RUDRA-JAYA</td><td rowspan="2" colspan="3">BHUDHARA</td><td rowspan="2">APAVATSA</td><td rowspan="2">APAVATSA</td><td>PARSANYA</td></tr>
<tr><td>SOSHA</td><td>JAYANTA</td></tr>
<tr><td>ASURA</td><td rowspan="3" colspan="2">MITRA</td><td rowspan="3" colspan="3">BRAHMA</td><td rowspan="3" colspan="2">ARYAKA</td><td>MAHENDRA</td></tr>
<tr><td>VARUNA</td><td>BHANU</td></tr>
<tr><td>PUSHPA -DANTA</td><td>SATYA</td></tr>
<tr><td>SUGRIVA</td><td rowspan="2">INDRA-JAYA</td><td rowspan="2">INDRA</td><td rowspan="2" colspan="3">VIVASVAT</td><td rowspan="2">SAVITRA</td><td rowspan="2">SAVITRA</td><td>BHRISA</td></tr>
<tr><td>DAUVARIKA</td><td>ANTA-RIKSHA</td></tr>
<tr><td>PITRA</td><td>MRISA</td><td>BHRINGA -RAJA</td><td>GANDHARVA</td><td>YAMA</td><td>GRIHA -KSHATA</td><td>VITATHA</td><td>PUSHAN</td><td>AGNI</td></tr>
</table>

Site plan: Param sayika plan—81 plots, radial

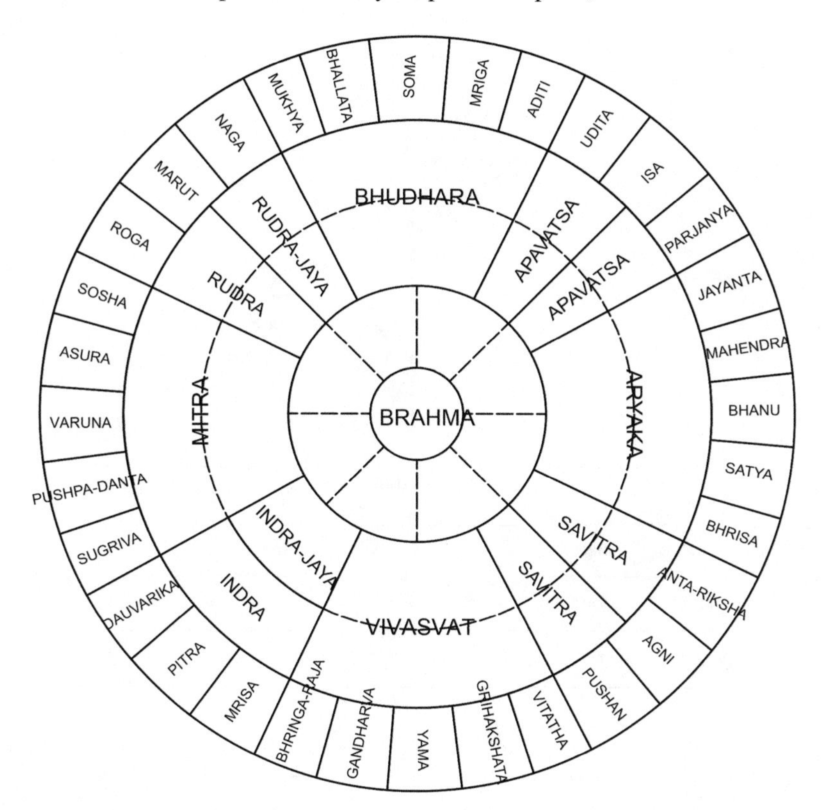

Site plan: Param sayika plan—81 plots, triangular

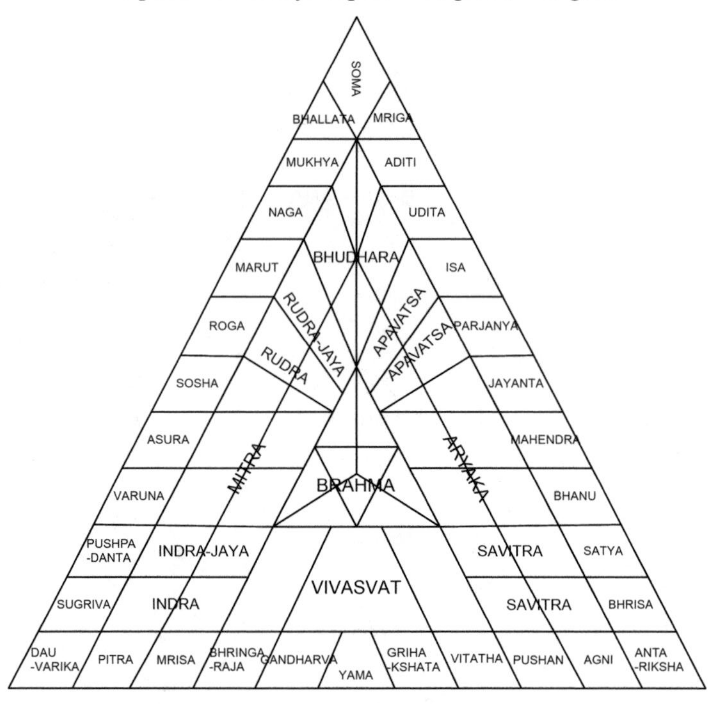

Site plan: Asana plan—100 plots

<table>
<tr><td>YAYO
ROGA</td><td colspan="2">NAGA</td><td>MUKHYA</td><td>BHALLATA</td><td>SOMA</td><td>MRIGA</td><td colspan="2">ADITI</td><td>UDITA
ISA</td></tr>
<tr><td rowspan="2">SOSHA</td><td colspan="2">RUDRAJAYA</td><td colspan="4" rowspan="2">BHUDHARA</td><td colspan="2">APAVATSA</td><td rowspan="2">PARJANYA</td></tr>
<tr><td colspan="2">RUDRA</td><td colspan="2">APAVATSA</td></tr>
<tr><td>ASURA</td><td colspan="2" rowspan="4">MITRA</td><td colspan="4" rowspan="4">BRAHMA</td><td colspan="2" rowspan="4">ARYAMAN</td><td>JAYANTA</td></tr>
<tr><td>VARUNA</td><td>MAHENDRA</td></tr>
<tr><td>PUSHPA
-DANTA</td><td>ADITYA</td></tr>
<tr><td>SUGRIVA</td><td>SATYA</td></tr>
<tr><td rowspan="2">DAUVARIKA</td><td colspan="2">INDRA-JAYA</td><td colspan="4" rowspan="2">VIVASVAT</td><td colspan="2">SAVITRA</td><td rowspan="2">BHRISA</td></tr>
<tr><td colspan="2">INDRA</td><td colspan="2">SAVITRA</td></tr>
<tr><td>PITRA
MRISA</td><td colspan="2">BHRINGA
-JAYA</td><td>GANDHAVA</td><td>YAMA</td><td>GRIHA
-KSHATA</td><td>VITATHA</td><td colspan="2">PUSHAN</td><td>ANTA
-RIKSHA
AGNI</td></tr>
</table>

Appendix III. Figure-Ground Abstraction of Temple to Read for Auspiciousness of Sites and Plot Shapes (Not to Scale)

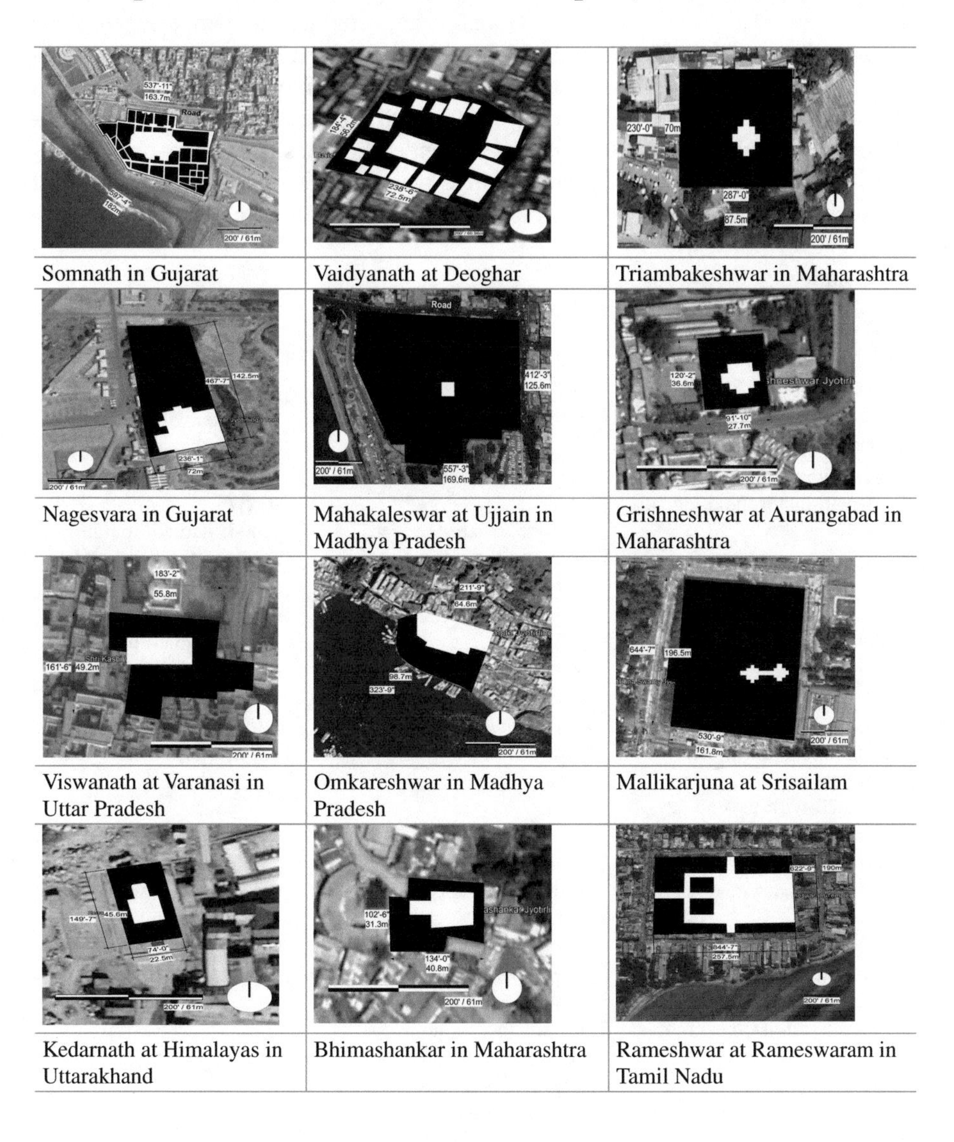

Somnath in Gujarat

Vaidyanath at Deoghar

Triambakeshwar in Maharashtra

Nagesvara in Gujarat

Mahakaleswar at Ujjain in Madhya Pradesh

Grishneshwar at Aurangabad in Maharashtra

Viswanath at Varanasi in Uttar Pradesh

Omkareshwar in Madhya Pradesh

Mallikarjuna at Srisailam

Kedarnath at Himalayas in Uttarakhand

Bhimashankar in Maharashtra

Rameshwar at Rameswaram in Tamil Nadu

Appendix IV. Internet References for Anecdotal History

Internet search was conducted using Google search engine, with temple legends and myths, as keywords through April of 2019. Number of results was noted. Key websites on the first page were recorded and reviewed. A detailed review of websites on the first page showed that even the websites with obtuse content were being captured under results. A deep dive into the content of the websites with relevant names revealed that the same or similar narrative was being repeated on websites and blogs, thus precluding the doubts about multiple narratives of one construct.

Kedarnath

Kedarnath Temple Legend: 316,000 results	Myths: 413,000 results	
Legend	Mythology	Repeat websites
https://en.wikipedia.org/wiki/Kedarnath_Temple	https://en.wikipedia.org/wiki/Kedarnath_Temple	https://en.wikipedia.org/wiki/Kedarnath_Temple
https://www.sacredyatra.com/kedarnath-history-and-legends.html	https://www.cntraveller.in/story/how-kedarnath-temple-survived-the-flood-and-400-years-under-ice/	https://www.cntraveller.in/story/how-kedarnath-temple-survived-the-flood-and-400-years-under-ice/
https://www.mahashivratri.org/kedarnath-temple-himalayas.html	https://www.speakingtree.in/allslides/hindu-mythology-of-shri-kedarnath	https://www.sacredyatra.com/kedarnath-history-and-legends.html
https://zeenews.india.com/entertainment/and-more/kedarnath-temple-legends-associated-with-the-divine_1954382.html	https://www.sacredyatra.com/kedarnath-history-and-legends.html	https://www.chardhamtour.in/kedarnath-history-and-legends.html
https://www.chardhamtour.in/kedarnath-history-and-legends.html	https://www.chardhamtour.in/kedarnath-history-and-legends.html	https://isha.sadhguru.org/us/en/wisdom/article/kedarnath-temple-crazy-cocktail-spirituality
https://www.cntraveller.in/story/how-kedarnath-temple-survived-the-flood-and-400-years-under-ice/	https://isha.sadhguru.org/us/en/wisdom/article/kedarnath-temple-crazy-cocktail-spirituality	
https://vedicfeed.com/interesting-facts-you-need-to-know-about-kedarnath-temple/	https://www.news18.com/news/india/kedarnath-shrine-was-under-snow-for-400-years-scientists-619076.html	
http://blessingsonthenet.com/indian-temple/article/859/legend-of-kedarnath-temple	http://www.kedarnath-dham.com/2012/01/history-of-kedarnath-temple.html	
https://isha.sadhguru.org/us/en/wisdom/article/kedarnath-temple-crazy-cocktail-spirituality		

Vishwanath

Vishwanath Temple Legend: 85,000 results	Myths: 330,000 results	
Legend	Mythology	Repeat websites
https://en.wikipedia.org/wiki/Kashi_Vishwanath_Temple	https://en.wikipedia.org/wiki/Kashi_Vishwanath_Temple	https://en.wikipedia.org/wiki/Kashi_Vishwanath_Temple
http://www.ramadajhvvns.com/blog/2018/02/05/what-is-the-story-behind-kashi-vishwanath-temple/	http://jyotirlingatemples.com/article/id/530/temple/43/mythology-of-kashi-vishwanath-temple	http://jyotirlingatemples.com/article/id/530/temple/43/mythology-of-kashi-vishwanath-temple
https://u4uvoice.com/the-legend-of-kashi-vishwanath-temple/	http://www.ramadajhvvns.com/blog/2018/02/05/what-is-the-story-behind-kashi-vishwanath-temple/	http://www.ramadajhvvns.com/blog/2018/02/05/what-is-the-story-behind-kashi-vishwanath-temple/
https://www.speakingtree.in/allslides/7-surprising-facts-about-kashi-vishwanath-temple	https://metrosaga.com/facts-about-kashi/	https://www.culturalindia.net/indian-temples/kashi-vishwanath.html
https://www.skyscanner.co.in/news/tips/kashi-vishwanath-temple	https://www.nativeplanet.com/travel-guide/7-interesting-facts-about-kashi-vishwanath-temple-002643.html	
https://www.mahashivratri.org/vishwanath-temple-varanasi.html	https://www.cultureholidays.com/Temples/kashi.htm	
http://jyotirlingatemples.com/article/id/530/temple/43/mythology-of-kashi-vishwanath-temple	https://www.quora.com/What-is-the-history-of-Kashi-and-Varanasi	
https://www.culturalindia.net/indian-temples/kashi-vishwanath.html	https://www.culturalindia.net/indian-temples/kashi-vishwanath.html	
https://topyaps.com/7-surprising-facts-kashi-vishwanath-temple/	https://www.shrikashivishwanath.org/mythology/importantparvas	
	http://ritsin.com/kashi-vishwanath-jyotirlinga-lord-shiva.html/	

Baijnath

Baba Baidyanath Temple **Legend:** 82,100 results	Myths: 217,000	
Legend	Mythology	Repeat websites
https://en.wikipedia.org/wiki/Baidyanath_Temple	https://en.wikipedia.org/wiki/Baidyanath_Temple	https://en.wikipedia.org/wiki/Baidyanath_Temple
http://www.babadham.org/history.php	https://www.explorebihar.in/the-history-of-baba-baidyanath-dham-temple-deoghar.html	https://www.myoksha.com/baidyanath-temple/
http://www.babadham.org/lingamorigin.php	https://trekkerpedia.com/culture/babadham-story-history-of-lord-shiva-temple-deoghar-jharkhand/	http://www.babadham.org/history.php
https://www.myoksha.com/baidyanath-temple/	https://www.myoksha.com/baidyanath-temple/	https://www.tripadvisor.com/ShowUserReviews-g1893710-d2697959-r210358142-Baba_Baidyanath_Temple-Deoghar_Deoghar_District_Jharkhand.html
https://www.tripadvisor.com/ShowUserReviews-g2531379-d4138801-r575980041-Baba_Basukinath_Dham-Dumka_Dumka_District_Jharkhand.html	https://deoghar.nic.in/history/	
https://www.tripadvisor.com/ShowUserReviews-g1893710-d2697959-r463104512-Baba_Baidyanath_Temple-Deoghar_Deoghar_District_Jharkhand.html	https://deoghar.nic.in/tourist-place/baidyanath-temple/	
https://www.templepurohit.com/hindu-temple/baidyanath-temple-jharkhand/	https://www.tripadvisor.com/ShowUserReviews-g1893710-d2697959-r210358142-Baba_Baidyanath_Temple-Deoghar_Deoghar_District_Jharkhand.html	
https://www.bharattemples.com/baidyanath-temple-baba-baidyanath-dham-deoghar-jharkhand/	http://babadham.org/about.php	
https://www.holidify.com/places/deoghar/baidyanath-dham-sightseeing-1734.html	http://www.babadham.org/history.php	
http://www.capertravelindia.com/jharkhand/baidyanath-dham.html	http://www.babadham.org/	

Somnath

Somnath Temple Legend: 251,000 results	Myths: 234,000 results	
Legend	Mythology	Repeat websites
https://www.speakingtree.in/blog/legend-of-somnath-temple	https://en.wikipedia.org/wiki/Somnath_temple	https://en.wikipedia.org/wiki/Somnath_temple
https://en.wikipedia.org/wiki/Somnath_temple	https://www.speakingtree.in/allslides/lesser-known-facts-about-somnath-temple	https://www.speakingtree.in/allslides/lesser-known-facts-about-somnath-temple
http://jyotirlingatemples.com/article/id/535/temple/44/mylthology%2D%2Dof-somnath%2D%2Dtemple	https://www.speakingtree.in/allslides/somnath-temple-the-history-of-neverending-glory	http://jyotirlingatemples.com/article/id/535/temple/44/mylthology%2D%2Dof-somnath%2D%2Dtemple
https://www.manishjaishree.com/2015/06/26/the-legend-of-somnath-temple/	http://jyotirlingatemples.com/article/id/535/temple/44/mylthology%2D%2Dof-somnath%2D%2Dtemple	https://www.tourmyindia.com/blog/interesting-facts-about-somnath-temple/
https://www.gujaratexpert.com/somnath-history/	https://www.tourmyindia.com/blog/interesting-facts-about-somnath-temple/	https://www.gujaratexpert.com/somnath-history/
https://jothishi.com/somnath-temple-all-about-gujarats-shrine-eternal/	https://www.gujaratexpert.com/somnath-history/	https://www.culturalindia.net/indian-temples/somnath-temple.html
https://www.tourmyindia.com/blog/interesting-facts-about-somnath-temple/	https://www.culturalindia.net/indian-temples/somnath-temple.html	
https://www.mahashivratri.org/somnath-temple-gujarat.html	https://www.indiapost.com/somnath-looted-destroyed-and-resurrected-17-times/	
https://www.culturalindia.net/indian-temples/somnath-temple.html	https://www.indianholiday.com/blog/somnath-temple-curse-of-the-moon/	
https://www.india.com/news-travel/do-you-know-this-unbelievable-legend-behind-gujarats-somnath-temple-3235330/	https://www.patheos.com/blogs/hindu2/2019/01/a-brief-history-of-the-somnath-temple/	

Nageshwar

Nageshwar temple Legend: 44,000 results	Myths: 48,400 results	
Legend	Mythology	Repeat websites
https://en.wikipedia.org/wiki/Nageshvara_Jyotirlinga	https://www.gujaratexpert.com/nageshwar-temple-history/	https://www.gujaratexpert.com/nageshwar-temple-history/
http://jyotirlingatemples.com/article/id/580/temple/51/legend-of-nageshwar-temple	https://en.wikipedia.org/wiki/Nageshvara_Jyotirlinga	https://en.wikipedia.org/wiki/Nageshvara_Jyotirlinga
https://www.gujaratexpert.com/nageshwar-temple-history/	http://ritsin.com/the-story-of-nageshwar-jyotirlinga-lord-shiva.html/	http://ritsin.com/the-story-of-nageshwar-jyotirlinga-lord-shiva.html/
https://www.gujarattourism.com/destination/details/10/312	https://www.gujarattourism.com/destination/details/10/312	https://www.gujarattourism.com/destination/details/10/312
http://blessingsonthenet.com/indian-temple/article/969/legend-of-nageshwar-temple	http://punetopune.com/conflicts-nageshwar-jyotirlinga/	https://www.mahashivratri.org/nageshwar-temple-dwarka.html
http://ritsin.com/the-story-of-nageshwar-jyotirlinga-lord-shiva.html/	https://www.mahashivratri.org/nageshwar-temple-dwarka.html	https://www.templepurohit.com/hindu-temple/nageshwar-jyotirling-dwarka/
	https://www.tripadvisor.com/ShowUserReviews-g660182-d3184202-r186347957-Dwarkadhish_Temple-Dwarka_Devbhumi_Dwarka_District_Gujarat.html	
https://www.mahashivratri.org/nageshwar-temple-dwarka.html	https://hinduism.stackexchange.com/questions/14359/what-is-the-story-of-nageshwar-jyotirlinga	
https://www.templepurohit.com/hindu-temple/nageshwar-jyotirling-dwarka/	https://www.templepurohit.com/hindu-temple/nageshwar-jyotirling-dwarka/	
https://myoksha.com/nageshwar-temple/		

Mahakaal

Mahakaleshwar Temple Legends: 90,400 results	Myths: 157,000 results	
Legend	Mythology	Repeat websites
https://www.mahashivratri.org/mahakaleshwara-temple-ujjain.html	http://jyotirlingatemples.com/article/id/546/temple/45/mythology-of-mahakaleshwar-temple	https://devdutt.com/articles/indian-mythology/temples-of-death.html
https://en.wikipedia.org/wiki/Mahakaleshwar_Jyotirlinga	https://devdutt.com/articles/indian-mythology/temples-of-death.html	https://en.wikipedia.org/wiki/Mahakaleshwar_Jyotirlinga
https://devdutt.com/articles/indian-mythology/temples-of-death.html	https://en.wikipedia.org/wiki/Mahakaleshwar_Jyotirlinga	https://www.speakingtree.in/blog/mahakaleshwar-temple-ujjain
http://blog.onlineprasad.com/legends-of-the-mahakaleshwar-jyotirlinga-temple/	https://daily.bhaskar.com/news/JM-secrets-of-thousands-of-years-old-mahakal-temple-no-one-knows-4341656-PHO.html	http://ritsin.com/mahakaleshwar-jyotirlinga-lord-shiva-as-mahakal.html/
https://www.speakingtree.in/blog/mahakaleshwar-temple-ujjain	https://www.quora.com/What-is-the-significance-of-the-mahakaleshwar-temple-in-Ujjain-I-am-planning-to-visit-but-need-more-clarity	https://www.tourmyindia.com/blog/unheard-facts-nagchandreshwar-temple/
https://www.tourmyindia.com/blog/unheard-facts-nagchandreshwar-temple/	https://www.speakingtree.in/blog/mahakaleshwar-temple-ujjain	
https://www.tourmyindia.com/blog/temples-in-india-with-interesting-untold-tales/	http://ritsin.com/mahakaleshwar-jyotirlinga-lord-shiva-as-mahakal.html/	
https://vedicfeed.com/mahakaleshwar-jyotirlinga-temple/	https://www.tourmyindia.com/blog/unheard-facts-nagchandreshwar-temple/	
http://hindutemples-india.blogspot.com/2019/03/mahakaleshwar-temple-ujjain-legends.html	https://www.culturalindia.net/indian-temples/mahakaleshwar-temple.html	
https://indroyc.com/2018/01/17/mahakaleshwar/	http://religions.iloveindia.com/indian-temples/mahakaleshwar-temple.html	

Bhimashankar

Bhimashankar Temple Legend: 90,400 results	Myths: 82,500 results	
Legend	Mythology	Repeat websites
https://jyotirlingatemples.com/article/id/575/temple/50/legend-of-bhimashankar-temple	https://www.artofliving.org/mahashivratri/bhimashankar-jyotirlinga	https://www.artofliving.org/mahashivratri/bhimashankar-jyotirlinga
https://www.speakingtree.in/blog/bhimashankar-jyotirling	https://blessingsonthenet.com/travel-india/destination/article/id/379/tour/id/343/history-of-bhimashankar	https://blessingsonthenet.com/travel-india/destination/article/id/379/tour/id/343/history-of-bhimashankar
https://blessingsonthenet.com/travel-india/destination/article/id/379/tour/id/343/history-of-bhimashankar	https://www.astroved.com/astropedia/en/temples/west-india/bhimashankar-jyotirlinga	https://www.astroved.com/astropedia/en/temples/west-india/bhimashankar-jyotirlinga
https://www.artofliving.org/mahashivratri/bhimashankar-jyotirlinga	https://jyotirlingatemples.com/article/id/575/temple/50/legend-of-bhimashankar-temple	http://ritsin.com/
https://www.astroved.com/astropedia/en/temples/west-india/bhimashankar-jyotirlinga	https://trendingnewswala.online/bhimashankar-temple/	
https://bhimashankar.in	https://pravase.co.in/thingstododetail/461/india/maharashtra/bhimashankar/bhimashankar-temple-one-of-twleve-jyotirlinga-history-importance-timing	
https://www.ttelangana.com/2020/09/bhimashankar-jyotirlinga-temple-maharashtra-full-details.html	https://en.wikipedia.org/wiki/Bhimashankar_Temple	
	https://ritsin.com/bhimashankar-jyotirlinga-lord-shiva-temple.html/	

Omkareshwar

Omkareshwar Temple Legend: 56,700 results	Myths: 72,200 results	
Legend	Mythology	Repeat websites
http://jyotirlingatemples.com/article/id/549/temple/46/-legend-of%2D%2Domkareshwar-temple	https://en.wikipedia.org/wiki/Omkareshwar_Temple	https://en.wikipedia.org/wiki/Omkareshwar_Temple
https://en.wikipedia.org/wiki/Omkareshwar_Temple	http://ritsin.com/omkareshwar-jyotirlinga-lord-shiva.html/	http://ritsin.com/omkareshwar-jyotirlinga-lord-shiva.html/
https://thecompletepilgrim.com/omkareshwar-temple/	http://jyotirlingatemples.com/article/id/550/temple/46/significance-of-omkareshwar%2D%2Dtemple	http://jyotirlingatemples.com/article/id/550/temple/46/significance-of-omkareshwar%2D%2Dtemple
https://www.indianholiday.com/blog/omkareshwar-jyotirlinga-legends-fall/	http://omkareshwar.org/history-of-the-omkareshwar/79/English	
http://ritsin.com/omkareshwar-jyotirlinga-lord-shiva.html/	https://blog.onlineprasad.com/omkareshwar-jyotirlinga-temple-history-and-importance/	
https://www.pinterest.com/pin/773845148441330100/	https://medium.com/@kartikjhori.ga/history-of-omkareshwar-jyotirlinga-temple-f79fc9c5a7d1	
https://www.pinterest.com/pin/589901251169318869/	https://www.thehindu.com/features/friday-review/history-and-culture/Myth-and-mystery/article16890757.ece	
http://hindutemples-india.blogspot.com/2019/04/Omkareshwar-temple-mandhata-madhya-pradesh.html	https://www.mahashivratri.org/omkareshwar-temple-madhya-pradesh.html	
	https://religions.iloveindia.com/indian-temples/omkareshwar-temple.html	
Other keyword searches: http://mudiraja.weebly.com/25-emperor-mandhatha-of-solar-race-ganga%2D%2D-mohenjo-daro-period.html https://pparihar.com/2016/02/01/king-mandhata-or-mandhatri/		

Triambakeshwar

Triambakeshwar Temple Legend: 47,800 results	Myths: 65,700 results	
Legend	Mythology	Repeat websites
https://www.mahashivratri.org/trimbakeshwar-temple-nasik.html	https://en.wikipedia.org/wiki/Trimbakeshwar_Shiva_Temple	https://en.wikipedia.org/wiki/Trimbakeshwar_Shiva_Temple
https://en.wikipedia.org/wiki/Trimbakeshwar_Shiva_Temple	https://www.speakingtree.in/blog/trimbakeshwar-jyotirlinga	http://ritsin.com/trimbakeshwar-jyotirlinga-lord-shiva.html/
http://blog.onlineprasad.com/the-legend-behind-the-trimbakeshwar-temple/	http://ritsin.com/trimbakeshwar-jyotirlinga-lord-shiva.html/	https://religions.iloveindia.com/indian-temples/trimbakeshwar-temple.html
http://ritsin.com/trimbakeshwar-jyotirlinga-lord-shiva.html/	https://www.tourmyindia.com/pilgrimage/tryambakeshwar.html	
https://www.outlookindia.com/traveller/ot-getaway-guides/trimbakeshwar-godavari-born/	https://vedicfeed.com/trimbakeshwar-shiva-temple/	
https://thecompletepilgrim.com/trimbakeshwar-shiva-temple/	https://www.india.com/news-travel/maha-shivratri-2018-interesting-facts-about-trimbakeshwar-jyotirlinga-temple-3226870/	
https://religions.iloveindia.com/indian-temples/trimbakeshwar-temple.html	https://www.holidify.com/places/nasik/trimbakeshwar-sightseeing-3063.html	
http://aurangabadelloraajanta.com/travel_related_services/nashik_trimbakeshwar.html	https://religions.iloveindia.com/indian-temples/trimbakeshwar-temple.html	
Other keyword searches: http://indiafacts.org/the-scientific-dating-of-the-ramayana/		

Rameshwaram

Rameshwaram Temple Legend: 111,000 results	Myths: 344,000 results	
Legend	Mythology	Repeat websites
http://jyotirlingatemples.com/article/id/554/temple/47/legend-of-rameswaram-temple/rameshwar-rameswaram	https://en.wikipedia.org/wiki/Rameswaram	https://en.wikipedia.org/wiki/Rameswaram
https://en.wikipedia.org/wiki/Ramanathaswamy_Temple	https://www.speakingtree.in/allslides/rameshwaram-temple-history	http://www.myrameswaram.com/tale-of-an-rameswaram-island
https://www.culturalindia.net/indian-temples/rameshwaram.html	http://www.myrameswaram.com/tale-of-an-rameswaram-island	https://www.ancient-origins.net/ancient-places-asia/ramanathaswamy-temple-and-its-infinite-corridors-009649
https://hindutemplesblog.wordpress.com/2017/10/12/history-of-rameshwaram-temple/	http://www.myrameswaram.com/history-of-rameswaram-temple	
https://www.ancient-origins.net/ancient-places-asia/ramanathaswamy-temple-and-its-infinite-corridors-009649	https://m.dailyhunt.in/news/india/english/east+coast+daily+eng-epaper-eeastco/ramanathaswamy+temple+rameshwaram+history+and+facts-newsid-87425553	
https://vedicfeed.com/facts-about-ramanathaswamy-temple/	https://www.ancient-origins.net/ancient-places-asia/ramanathaswamy-temple-and-its-infinite-corridors-009649	
http://blessingsonthenet.com/indian-temple/article/584/legend-of%2D%2Drameswaram%2D%2Dtemple	http://www.rameswaramtemple.tnhrce.in/history-rameswaram.html	
https://www.mahashivratri.org/rameshwaram-temple-tamil-nadu.html		
http://www.myrameswaram.com/tale-of-an-rameswaram-island		
http://jyotirlingatemples.com/article/id/554/temple/47/legend-of-rameswaram-temple/rameshwar-rameswaram		

Ghrishneshwar

Ghrishneshwar Temple legends 39,000 results	Ghrishneshwar Temple myths 8560 results	
Legend	Mythology	Repeat websites
https://en.wikipedia.org › wiki › Grishneshwar_Temple	https://en.wikipedia.org › wiki › Grishneshwar_Temple	https://en.wikipedia.org › wiki› Grishneshwar_Temple
https://www.artofliving.org/mahashivratri/ghrishneshwar-jyotirlinga	https://jyotirlingatemples.com/article/id/592/temple/53/story-of-grishneshwar-temple	http://ritsin.com/grishneshwar-jyotirlinga.html/
https://jyotirlingatemples.com/article/id/590/temple/53/legend-of-grishneshwar-temple	http://ritsin.com/grishneshwar-jyotirlinga.html/	
https://www.astroved.com/astropedia/en/temples/west-india/grishneshwar-jyotirlinga	https://aastik.in/ghushmeshwar-jyotirlinga-temple-story/	
http://ritsin.com/grishneshwar-jyotirlinga.html/	https://www.youtube.com/watch?v=Uly_8sYdwZs	
https://www.hindu-blog.com/2019/05/grishneshwar-temple-history-and.html/	https://timesofindia.indiatimes.com/city/aurangabad/women-more-equal-than-men-at-this-temple/articleshow/51722012.cms	
https://myoksha.com/grishneshwar-temple/	https://behindeverytemple.org/hindu-temples/shiva/grishneshwar-temple/	
https://www.tirthyatraindia.com/12-jyotirling-ghrishneshwar-temple.html	https://english.newsnationtv.com/lifestyle/health-and-fitness/maha-shivaratri-2020-all-you-must-know-about-mahadevs-12-jyotirlingas-and-their-significance-254353.html	
http://hindutemples-india.blogspot.com/2019/03/mahakaleshwar-temple-ujjain-legends.html	https://www.culturalindia.net/indian-temples/mahakaleshwar-temple.html	
https://indroyc.com/2018/01/17/mahakaleshwar/	http://religions.iloveindia.com/indian-temples/mahakaleshwar-temple.html	
http://jyotirlingatemples.com/article/id/590/temple/53/legend-of-grishneshwar-temple		

Mallikarjuna Srisailam

Mallikarjuna Srisailam Temple legends 31,200 results		
Legend	Mythology	Repeat websites
http://jyotirlingatemples.com/article/id/559/temple/48/legend-of-srisailam%2D%2Dtemple	https://en.wikipedia.org/wiki/Mallikarjuna_Jyotirlinga	https://en.wikipedia.org/wiki/Mallikarjuna_Jyotirlinga
https://en.wikipedia.org/wiki/Mallikarjuna_Jyotirlinga	http://www.srisailamonline.com/history.html	http://jyotirlingatemples.com/article/id/559/temple/48/legend-of-srisailam%2D%2Dtemple
https://chaibisket.com/srisailam-temple-legends/	http://jyotirlingatemples.com/article/id/559/temple/48/legend-of-srisailam%2D%2Dtemple	
http://www.srisailamonline.com/legends.html	https://shaivam.org/temples-special/12-jyotirlingas-2-srisailam-temple-sri-mallikarjuna-jyothirlingam	
https://sacredsites.com/asia/india/srisailam.html	https://tms.ap.gov.in/SSLBMS/cnt/History	
http://ritsin.com/jyotirlinga-mallikarjuna-srisailam-temple.html/	http://ritsin.com/mallikarjun-jyotirlinga-lord_shiva-lord_kartikeya.html/	
http://jyotirlingatemples.com/article/id/559/temple/48/legend-of-srisailam%2D%2Dtemple	https://gotirupati.com/srisailam-temple/	
	https://www.astroved.com/astropedia/en/temples/south-india/sri-mallikarjuna-swamy-temple	
	https://thecompletepilgrim.com/mallikarjuna-swamy-temple/	

References

Acharya, P. K. (1927a). *Manasara Series: Vol. 1 – A Dictionary of Hindu Architecture*. Oxford: Oxford University Press, reprinted in 1934, 1995, 1997, 2008, 2015, Low Price Publications.
Acharya, P. K. (1927b). *Manasara Series: Vol. 2 – Indian Architecture According to Manasara-Silpasastra*. Oxford: Oxford University Press, reprinted in 1995, 2004, 2011, Low Price Publications.
Acharya, P. K. (1934a). *Manasara Series: Vol. 3 – Manasara – Sanskrit Text with Critical Notes*. Oxford: Oxford University Press, reprinted in 1995, 2011, Low Price Publications.
Acharya, P. K. (1934b). *Manasara Series: Vol. 4 – Architecture of Manasara*. Oxford: Oxford University Press. Translation in English, reprinted in 1995, 1998, 2006, 2015, Low Price Publications.
Acharya, P. K. (1934c). *Manasara Series: Vol. 5 – Architecture of Manasara – Illustration of Architectural and Sculptural Object*. Oxford: Oxford University Press, reprinted in 1995, 1998, 2006, 2013, Low Price Publications.
Acharya, P. K. (1946). *Manasara Series: Vol. 7 – An Encyclopaedia of Hindu Architecture*. Oxford: Oxford University Press, reprinted in 1995, 2000, 2010, Low Price Publications.
Alexander, C., Ishikawa, S., and Silverstein, M. (1977). *A Pattern Language: Towns, Buildings, Construction*. Oxford: Oxford University Press.
Araabi, H. F. (2016). A Typology of Urban Design Theories and Its Application to the Shared Body of Knowledge. *Urban Design International*, *21*(1), 11–24.
Balgooyen, T. G. (1973, November). Toward a More Operational Definition of Ecology. *Ecology*, 54(6), 1199–1200. Wiley on behalf of the Ecological Society of America.
Ben-Joseph, E., and Gordon, D. (2000). Hexagonal Planning in Theory and Practice. *Journal of Urban Design*, 5(3), 237–265.
Bharne, V., and Krusche, K. (2012). *Rediscovering the Hindu Temple: The Sacred Architecture and Urbanism of India*. Newcastle upon Tyne: Cambridge Scholars Publishing.
Bhatt, R. (2001). Indianizing Indian Architecture: A Postmodern Tradition. *Traditional Dwellings and Settlements Review*, *13*(1), 43–51.
Borelli, S., Conigliaro, M., Quaglia, S., and Salbitano, F. (2017). Urban and Peri-urban Agroforestry as Multifunctional Land Use. In *Agroforestry*. Singapore: Springer, pp. 705–724.
Brown, P. (1959). Indian Architecture: Buddhist and Hindu Period. In *Encyclopedia of Indian Temple Architecture: Vol. 2, Part 2, North India – Period of Early Maturity*. Bombay: Taraporevala.
Calvino, I. (1974). *Invisible Cities* (1st Harvest/HBJ edn). New York: Harcourt Brace Jovanovich.
Chakrabarti, V. (1998). Indian Architectural Theory: Contemporary Uses of Vastu Vidya. *Journal of the Royal Asiatic Society*, 11(2) (2001), 253–324.

A. Sharma, *Mandala Urbanism, Landscape, and Ecology*,
https://doi.org/10.1007/978-3-030-87285-4

Chakrabarti, V. (2013). *Indian Architectural Theory and Practice: Contemporary Uses of Vastu Vidya*. London: Routledge.
Ching, F. D. K. (1995). *A Visual Dictionary of Architecture*. New York: Wiley.
Duany, A., and Plater-Zyberk, E. (2006a). *Towns and Town-Making Principles*. New York: Rizzoli Publications.
Duany, A., and Plater-Zyberk, E. (2006b, 2021). *The New Urbanism*. Last ret 2021b 11 16. Mumbai: The Idea Studio.
Deva, K. (1996). Pan-Indian Style: North India, c. 250 B.C.–A.D. 400. In Dhaky, M. A. (Ed), *Encyclopaedia of Indian Temple Architecture Vol. 2 Part 1 Text North India Foundations of North Indian Style.* Princeton: Princeton University Press, pp. 3–18.
Dhaky, M. A. (2005). *Indian Temple Traceries*. Gurgaon: American Institute of Indian Studies; New Delhi: D.K.
Egerton, F. N. (2013, July). History of Ecological Sciences. Part 47: Ernst Haeckel's Ecology. *Bulletin of the Ecological Society of America*, 94(3), 222–244. Wiley on behalf of the Ecological Society of America.
Farr, D. (2007, 2011). *Sustainable Urbanism: Urban Design with Nature*. New York: Wiley.
Fergusson, J., Burgess, J., Spiers, R. P. (1910). *History of Indian and Eastern Architecture* (Vol. II). London: John Murray Publishers.
Friederichs, K. (1958, January). A Definition of Ecology and Some Thoughts About Basic Concepts. *Ecology*, 39(1), 154–159. Wiley on behalf of the Ecological Society of America.
Gandotra, A. (2011). *Indian Temple Architecture: Analysis of Plans, Elevations, and Roof Forms* (Vols. 1, 2, 3). New Delhi: Shubhi Publications.
Haeckel, E. (1866). *Generelle Morphologie der Organismen. Erster Band: Allgemeine Anatomie der Organismen*. Berlin: Reimer, S. IXXXII, 1574.
Hardy, A. (2007). *The Temple Architecture of India*. London: Wiley.
Hardy, A. (2012). Indian Temple Typologies. In Lorenzetti, T. and Scialpi, F. (Ed), *Glimpses of Indian History and Art Reflections on the Past, Perspectives for the Future. Proceedings of the International Congress Rome*, 18–19 April 2011, Sapienza Università Editrice, Roma.
Hardy, A. (2016). Hindu Temples and the Emanating Cosmos. *Religion and the Arts*, 20(1–2), 112–134.
Jain, P. (2019). Climate Engineering from Hindu-Jain Perspectives. *Zygon*, 54(4), 826.
Jain, P. (2011). *Dharma and Ecology of Hindu Communities: Sustenance and Sustainability*. Farnham/Burlington: Ashgate.
Jain, B. (1936). *Vaastusaar Prakaran, Jain Viividh Granthmala*. Ajmer: The Diamond Jubilee Press.
Jain, S. (2005). Vaishnav Havelis in Rajasthan-Origin and Continuity of a Temple Typology. In *Biennial Conference of the European Association of South Asian Archaeologists*, London, UK.
Kelbaugh, D. (2015). The Environmental Paradox of the City, Landscape, Urbanism, and New Urbanism. *Consilience*, 13, 1–15.
Kramrisch, S. (1976). *The Hindu Temple – Vol. 2: The Images of the Temple*. Delhi: Motilal Banarasidass, first published by the University of Calcutta, 1946.
Krier, L. (2009). *The Architecture of Community*. Washington, DC: Island Press.
Krier, L. (1984, July/August). Urban Components. *Architectural Design*, 54, 43–49.
Krier, L. (2010). Growth: Maturity or Over-Development. In Prashad, D. (Ed.), *New Architecture and Urbanism – Development of Indian Traditions*. Cambridge: Cambridge Scholars Publishing.
Lee, S., & Kim, Y. (2021). A Framework of Biophilic Urbanism for Improving Climate Change Adaptability in Urban Environments. *Urban Forestry & Urban Greening*, 61, 127104.
Lorenzetti, T. (2015). *Understanding the Hindu Temple*. Berlin: EB-Verlag.
Lorenzetti, T., and Scialpi, F. (Eds.). (2012). *Glimpses of Indian History and Art: Reflections on the Past, Perspectives for the Future; Proceedings of the International Congresses, Rome, 18–19 April 2011*. Sapienza Università Ed.
Lynch, K. (1960). *The Image of the City*. Cambridge: MIT Press.

Luca, O. (2017). Theoretical and Empirical Researches in Urban Management. *Considerations on Climate Strategies and Urban Planning: Bucharest Case Study*, 1–8.

Mehaffy, M. W., and Tigran, H. (2020). New Urbanism in the New Urban Agenda: Threads of an unfinished reformation. *Urban Planning*, 441-452.

Meister, M. W. (1981). Forest and Cave: Temples at Candrabhāgā and Kansuān. *Archives of Asian Art*, 34, 56–73. University of Hawai'i Press.

Meister, M. W., and Dhaky, M. A. (1992). *Encyclopedia of Indian Temple Architecture: Vol. 2, Part 2, North India – Period of Early Maturity: North India: Period of Early Maturity C.AD700-900*. New York: Oxford University Press.

Meister, M. W., and Dhaky, M. A. (1999). *Encyclopaedia of Indian Temple Architecture. South India. Lower Dravidadesa. 200 B. C.–A. D. 1324*. New Delhi: Manohar Publishers.

Meister, M. W. (2006). Mountain Temples and Temple-Mountains: Masrur. *Journal of the Society of Architectural Historians*, 65(1), 26–49.

Morley, I., and Renfrew, C. (Eds). (2010). *Archeology of Measurement: Comprehending Heaven, Earth and Time in Ancient Societies*. Cambridge/New York: Cambridge University Press.

Mookerji, R. K. (1960). *Chandragupta Maurya and His Times*. First edition published in 1940. Retrieved from Archive.Org

Moudon, A. V. (1994). Getting to Know the Built Landscape: Typomorphology. In *Ordering Space: Types in Architecture and Design*. New York: Van Nostrand Reinhold, pp. 289–311.

Narayanan, V. (2001). Water, Wood, and Wisdom: Ecological Perspectives from the Hindu Traditions. *Daedalus*, 130(4), 179–206.

Neubauer, J. (1981). The Stepwells of Gujarat: An Art-Historical Perspective. New Delhi: Abhinav Publications.

Raddock, E. (2017). Choosing an ācārya for Temple Construction and Image Installation. In *Consecration Rituals in South Asia*. Leiden: Brill, pp. 198–222.

Raddock, E. E. (2011). *Listen How the Wise One Begins Construction of a House for Viṣṇu: vijānatā yathārabhyaṃ gṛhaṃ vaiṣṇavaṃ śṛṇv evaṃ Chapters 1–14 of the Hayaśīrṣa Pañcarātra* (Doctoral dissertation, University of California, Berkeley).

Radhakrishnan, S. (1994). *The Principal Upanishads*. Indus. New Delhi: An Imprint of Harper Collins Publishers. First published in 1953 reprinted: 1978, and republished: 1995.

Radhakrishnan, S. (2008). *Indian Philosophy – Volume 1* (2nd edn). New Delhi: Oxford University Press. First published in 1923 and subsequently by various publishers in 1929, 1940, 1989, 1996, 2008, 2010.

Rademacher, A. (2018). *Consciousness and Indianness: Making Design "Good." Building Green: Environmental Architects and the Struggle for Sustainability in Mumbai*. Berkeley: University of California Press, pp. 108–132.

Rajan, K. V. S. (1972). *Indian Temple Styles: The Personality of Hindu Architecture*. New Delhi: Munshiram Manoharlal.

Rao, S. K. R., and Vikhanasacharyulu, D. (1997). *Devalaya-Vastu* (Vol. 1). Bangalore: Kalpatharu Research Academy.

Rising, H. H. (2015). *Water Urbanism: Building More Coherent Cities* (Doctor of Philosophy Dissertation). Department of Landscape Architecture, University of Oregon. Advisors: Ribe, R., Lobben, A., Ruggeri, D. Berkman, E.

Seidler, R., and Bawa, K. S. (2016). Ecology. In Adamson, J. Gleason, W. A., Pellow, D. N. (Eds.), *Keywords for Environmental Studies*. New York: NYU Press, pp. 71–75.

Sharma, U. M. D. (1936). *Sankhyateertha: Sutradharamandana: Devtamurtiprakaranam and rupamandanam*, Sanskrit. Calcutta: Calcutta Sanskrit Granthmala-12, Metropolitan Printing and Publishing House Limited.

Sastri, G. T. (1919). *The Mayamata of Mayamuni* (Trivandrum Sanskrit Series, No. LXV). Travancore: The Superintendent, Government Press, Government of His Highness the Maharajah of Travancore.

Sastri, G. T. (1913). *The Vastuvidya* (No. XXX). Travancore: The Travancore Government Press, Travancore, Government of His Highness the Maharajah of Travancore.

Schneider, W. J., David, A. R., and Andrew, M. S. (1973). *Role of Water in Urban Planning and Management* (Vol. 1, U.S. Geological Survey Circular 60). Washington, DC: United States Geological Survey.

Sharma, A. (2015, March). Urban Greenways: Operationalizing Design Syntax and Integrating Mathematics and Science in Design. *Frontiers of Architectural Research*, 4(1), 24–34. Elsevier B.V.

Sharma, S. D. (2008). *Vaastu Sarvasya*. New Delhi: Ranjan Publications.

Shukla, D. N. (1967). *Samrangana Sutradhara Part II: Royal Palace and Royal Arts*. Lucknow: Vastuvanmaya Prakasana-sala.

Shukla, D. N. (1961). *Hindu Science of Architecture*. Gorakhpur: Gorakhpur University.

Singh, R. P. B. (2011). Sacred Geography and Cosmic Geometries: Interfaces in Holy Places of North India and Link to Leonardo da Vinci's Images. *Asiatica Ambrosiana saggi e ricerche di cultura religioni e societ dell'Asia* [Accademia Ambrosiana, Piazza Pio XI, 2 – 23123 Milano, Italy], nr. 3: pp. 31–82! (Proceedings of the Dies Academicus 2010: Seminar on Geography & Cosmology Interfaces in Asia and Europe: 22–23 October 2010).

Sinha, A. (1996). Architectural Invention in Sacred Structures: The Case of Vesara Temples of Southern India. *Journal of the Society of Architectural Historians*, 55(4), 382–399.

Sinha, A. (2000). *Imagining Architects: Creativity in the Religious Monuments of India*. Newark: University of Delaware Press.

Sinha, A. (2006). *Landscapes in India: Forms and Meanings*. Boulder: University Press of Colorado.

Sinha, A. (1998). Design of Settlements in the Vaastu Shastras. *Journal of Cultural Geography*, 17(2), 27–41.

Sompura, P. O., and Sutradhar, N. (2011, 1965). *Prasadamanjari: Vikram Samwat* 2021 1965, Ahmedabad: Shri Balvantrai P. Sompura and Brothers, p. 27.

Sompura, P. O. (1975). The Vastuvidya of Visvakarma. In Chandra, P. M. (Ed). *Studies in Indian Temple Architecture*: Papers presented at a Seminar held in Varanasi, 1967. Gurgaon: American Institute of Indian Studies.

Sowell, J., and Wiedemann, N. (2009). Sponge Urbanism: The Cellular Redevelopment of New Orleans. *Journal of Architectural Education (1984-)*, 62(4), 24–31.

Srikanth, N., Tewari, D., Mangal, A. K., et al. (2015). *World Journal of Pharmacy and Pharmaceutical Sciences the Science of Plant Life (vriksha ayurveda) in Archaic Literature: An Insight on Botanical, Agricultural and Horticultural Aspects of Ancient India*. New Delhi: Central Council for Research in Ayurvedic Sciences, Ministry of Ayush, Government of India.

Subba Rao, D. V., Srinivasa Rao, K., Iyer, C. S. P., and Chittibabu, P. (2008). Possible Ecological Consequences from the Sethu Samudram Canal Project, India. *Marine Pollution Bulletin*, 56(2), 170–186.

Tarr, G. (1970). Chronology and Development of the Chāḷukya Cave Temples. *Ars Orientalis*, 8, 155–184.

The Times of India (TOI). (2021, September 11). *Uttar Pradesh; ASI Unearths 1500 Years Old Temple of Gupta Era, with Inscriptions in Shankha Lipi*. https://www.opindia.com/2021/09/asi-team-unearths-remains-of-1500-year-old-temple-from-Gupta-era/

Tillotson, G. (2005). The Indian Temple Traceries. By MA Dhaky. pp. xix, 490, figs. 55, plates 348. New Delhi, American Institute of Indian Studies & DK Printworld (P) Ltd. *Journal of the Royal Asiatic Society*, 15(3), 369–371.

Tillotson, G. H. R. (2014). *Paradigms of Indian Architecture: Space and Time in Representation And design*. London: Routledge.

Thadhani, D., and Duany, A. (2010). *The Language of Towns & Cities: A Visual Dictionary* (1st edn). New York: Rizzoli.

Udayakumar, S. (1995). A Study of Art and Architecture. *Scenographic Fashion Shows Alfonso Cuaron's Cinematography and the Mexican Culture The Art of Capital Art, Politics and*, 101.

Vandyck, F., & Bertels, I. (2017). Typology & Mixity: An Approach to Retrofit Production in the City? *Acta Technica Napocensis: Civil Engineering & Architecture*, 60(3), 77–87.

Van der Klaauw, C. J., and Meyer, A. (1936). Ökologische Studien und Kritiken: II. Zur Geschichte der Definitionen der Ökologie, besonders auf Grund der Systeme der zoologischen Disziplinen, Sudhoffs Archiv für Geschichte der Medizin und der Naturwissenschaften, Bd. 29, H. 3 (Oktober 1936), pp. 136–177.

Vatsyayan, K. (1986). Vastu-Purusha Mandala. In Kagal, C. (Ed.), *Vistāra – The Architecture of India, Catalogue of the Exhibition*. The Festival of India, 1986, in Architexturez South Asia.

Williams, K., and Ostwald, M. J. (2015). *Architecture and Mathematics from Antiquity to the Future*. Cham: Birkhäuser.

Index

A. Sharma, *Mandala Urbanism, Landscape, and Ecology*,
https://doi.org/10.1007/978-3-030-87285-4

Printed in the United States
by Baker & Taylor Publisher Services